Motorbooks International Illustra

Illustrated DODGE BUYER'S ★ GUIDE™

Richard M. Langworth

For Lee Iacocca, who gave Dodge a future,
and Robert Lutz, who ensured it.

First published in 1995 by Motorbooks International Publishers & Wholesalers, PO Box 2, 729 Prospect Avenue, Osceola, WI 54020 USA

Library of Congress Cataloging-in-Publication Data
Langworth, Richard M.
Illustrated Dodge buyer's guide/by Richard Langworth.
p. cm.—(Motorbooks International illustrated buyer's guide series)
Includes index.
ISBN 0-7603-0064-X (pbk.)
1. Dodge automobile—Purchasing. 2. Dodge automobile—History. I. Title. II. Series.
TL215.D6L36 1995
629.222'2—dc20 95-24760

On the front cover; This immaculate 1969 Charger Daytona is owned by Bill and Barbara Jacobsen of Silver Dollar Classics in Odessa, Florida. *Mike Mueller*

On the back cover: The voluptuous 1995 Viper takes the title as Cobra of the '90s; Dodge's top-line product for 1961 was the big Polara.

Printed and bound in the United States of America

Contents

Acknowledgments

I should like to thank Brad Rosenbush of the Chrysler Historical Collection for his diligent work in locating some seventy-five of the photos herein which I did not have in my own collection, and which were indispensable for the comprehensive illustrations on which the Buyer's Guide series relies. I also thank Jane Mausser, the editorial staff of Motorbooks International, and the countless Dodge owners, drivers, and collectors who provided information or experiences from which to draw upon. I acknowledge the *Old Cars Value Guide* for prices of most "condition 1" cars in the 1982–1995 price history sections. Finally, I recall again with grateful thanks the comments of many of those on the Dodge scene when some of these cars were new, including the late Virgil Exner, the late Murray Baldwin, the late Herb Weissinger, and the very much alive Jeff Godshall, whose distinguished career at Chrysler Design makes him a maker as well as a writer of automotive history, one of the few to do both, and to do both well.

Introduction

Dodge was the first car brand of which I became conscious. This occurred in 1949 when my father bought the first new car our family had ever owned, a 1949 gray Meadowbrook sedan, from Schick Motors, a Dodge-Plymouth dealership on Staten Island, New York. The Meadowbrook, named for a palatial Detroit estate, was anything but deluxe, being a detrimmed version of the high-end Coronet with rubber floor mats and plain striped upholstery (which was immediately slipcovered, the current fashion). It also lacked Fluid Drive, Dodge's famous semi-automatic, at which my father balked in favor of a conventional three-speed column shift. But I admired its charging ram hood mascot, the mahogany-grained dashboard with jewel-like square instruments, the rich-sounding radio with Chrysler's patented sound control that gave you a range from "voice" to "mello," and three-way-visible taillights the color of Charms' cherry lollipops. My father called it the "gray ghost." In fact it was anything but ghostlike, a product of the K. T. Keller school of design nicknamed "three-box styling": one box piled on top of two others, as if it had been built by a bricklayer. (Indeed, Virgil Exner, the stylist who put Chrysler design at the head of the class by 1957, later suggested to me that a bricklayer had designed *all* the '49s. . . .)

The second Dodge I clearly remember was a 1955 Coronet Lancer owned by Ronnie Ferger's parents. I used to ogle it in its garage at 130 Dongan Street. We were still too young to drive, but I can remember peering into the Ferger Coronet with marvel in my eyes and love in my heart. I'd always liked Chrysler products. (In those days if your parents bought Chrysler or GM or Ford, you liked the same company and put its product logos on your bike; this was a spiritual thing, passed on from parents to progeny.) But this '55 was different, resembling no Dodge I'd known before. To begin with, there was the paint job: *three-tone paint!* Ferger's was a typical combination: pink, gray, and cream, beautifully glossy, the colors separated by slim chrome borders. Three-toning was a "first" for Dodge that year, along with DeSoto's Fireflite Coronado and Packard's Caribbean, but the latter two were limited editions, and as a kid I don't remember ever seeing one new. Dodge Coronets, Royals, and Custom Royals, on the other hand, were everywhere.

In retrospect we look upon the three-tone paint job as a prime example of fifties decadence, but I affirm that on the '55 Dodge it looked pretty good. A prominent, wide hood bulge was edged in chrome strips which tapered back along the beltline, dipping briefly under the rear windows and carrying smoothly onto the tail. The gray was daubed in his area, then back across the deck. Underneath was the pink, and the cream was on the roof. I guess it proved too much, since Mr. Ferger later had the roof repainted pink to match the lower body, ending up with a conventional two-tone.

The dashboard was unforgettable—space age stuff in 1955. Gone was the symmetrical look of the recent past, and the fake mahogany of Dad's old Meadowbrook (since traded in for a '52 Chrysler Windsor). The '55 Dodge dash was a big, concave oval, sweeping in front of the driver and halfway across to the passenger's side, where it fell away, the space on the right occupied by a glovebox. The metal around it was painted a nonreflective color, with a texture that looked like velvet. Facing the driver was a large speedometer, flanked by the round minor instruments, with the control knobs on a rounded bank underneath: logical, easy to use, and sharp as all get-out.

Another thing that caught my youthful eye through the wraparound windshield was the shift lever, or rather the lack of one. Where did it go? Down on the dashboard, where a "Flight Control" wand sprouted to control the PowerFlite automatic transmission through a mechanical linkage. Far out! My amazement doubled the following year when the shifter vanished utterly, replaced by a pod full of "Magic Touch" pushbuttons mounted where you'd least expect it, to the left of the steering wheel. Also mechanically controlled, the pushbuttons were popular, lasting on various Chrysler products into the mid-1960s.

Later I met the man who designed that dash, and most of the car itself: Murray Baldwin, a colleague of Exner's. I became hooked on the story of Exner's years at Chrysler, how he took a company renowned for the dowdy plainness of its styling and gave it products that were the envy of the design profession by 1956 (the year of the "Forward Look") and 1957 (the year of "Flight Sweep").

I still like Dodges. The latest new one of which I am conscious is the red Stealth R/T driven by my dentist, undoubtedly secured, as I keep reminding him, with the proceeds of crown work on what's left of my teeth. In the barn as I write is a 1984 Dodge Rampage, a cute little pickup truck that I keep trying to convince myself is a genuine collectible, which I admire so much that I sneaked it into this Buyer's Guide about Dodge cars, on the grounds that it is more car than truck, or at least half-car, or something like that. Rampage is, by the way, the smallest Dodge truck ever built; I have a friend who owns the largest, a Big Horn tractor trailer, and it is our ambition someday to do a comparison road test and a drag-race.

Dodge has a distinguished pedigree. It was born of hard-drinking, Ford-hating John and Horace Dodge, raised to adulthood by Walter Percy Chrysler, and given a performance image in the fifties by the Red Ram V-8 and in the sixties by the Hemi. It was this early pretense of sportiness that gradually molded Dodge into the "performance marque" of the company, attracting a youthful following that it had never had in the days of "three-box styling" and the Meadowbrook.

After a close call with cancer in the 1970s, Dodge was rescued by Dr. Iacocca in the 1980s, and today builds some of the most exciting cars on the world's roads. I say "world's" intentionally. Dodge is no longer an ill-built, over-bodied, under-braked, fuel-thirsty land yacht. It's a world-class competitor. There is not a red-blooded car fan in England or Germany or Japan who doesn't admire the speed and stamina of the Viper, the svelte looks of the Stealth, the dynamic beauty of the Intrepid. These modern Dodges are worthy successors to the Challengers, R/Ts, Coronets, and D-500s of the past.

In evaluating the Dodge of today most commentators seem highly positive. It is the only Division still intact from the time when Walter Chrysler set out to mimic General Motors with a compartmentalized corporate structure meshed to a pecking order which the typical customer would mount, a step at a time. Since 1930 there had been a Chrysler, DeSoto, and Plymouth division as well as Dodge, with each dealership getting Plymouth, for volume, along with one of the other three. The idea was that you'd start with a Plymouth, then trade for a Dodge, DeSoto, and finally a Chrysler as your worldly goods accumulated.

For thirty years this worked just fine, but in 1960 following a disastrous recession, DeSoto was blended into Plymouth-Valiant Division, and then vanished forever; soon Chrysler and Plymouth were merged, while the shortlived separate-make Imperial melded back into Chrysler and eventually faded away.

Through all this, Dodge Division survived intact.

There are ample reasons why Dodge has retained its distinct identity, and many of them are in this book. From the days when John and Horace Dodge set out to build a better Model T to the approaching end of this century, the products of this division have attracted more than their mathematical share of fans and interest. Vast numbers of old Dodges have survived, to be restored lovingly by a new generation who never knew the cars when new. The aim of this book is to codify and categorize the collectible Dodge for that new generation of car collectors.

This Buyer's Guide covers every Dodge built from 1914 through the mid-1970s, and collectible (or potentially collectible) Dodges from the mid-1970s to the present, right on through the Stealth and Viper. It also includes those highly desirable recent Dodges which have benefited from the hand of Carroll Shelby, bearing both Shelby and Dodge marque names.

It is, of course, impossible to predict now how a car as new as the Viper will be viewed by the collector crowd twenty or thirty years hence, but you can get a rough idea by going out and driving one. It's a nice thing that in an age of jellybean-shaped transportation modules, some companies with distinguished names are still building automobiles to excite the senses—cars that stand to be remembered long after their time.

Each chapter adheres to the following format.

History: a background, placing the models covered in historical perspective, and noting their success or failure at the time.

Identification: how to tell one model from another, and features of various model years.

Appraisal: peculiar maladies, driveability problems and trouble areas to look at before you buy a particular model, including problems peculiar to all examples and problems likely to develop on many of them.

Summary and Prospects: current collectibility and/or prospects for changes in this status.

Price History: a thirteen-year comparison of top values for near-perfect show-quality examples of various cars under discussion.

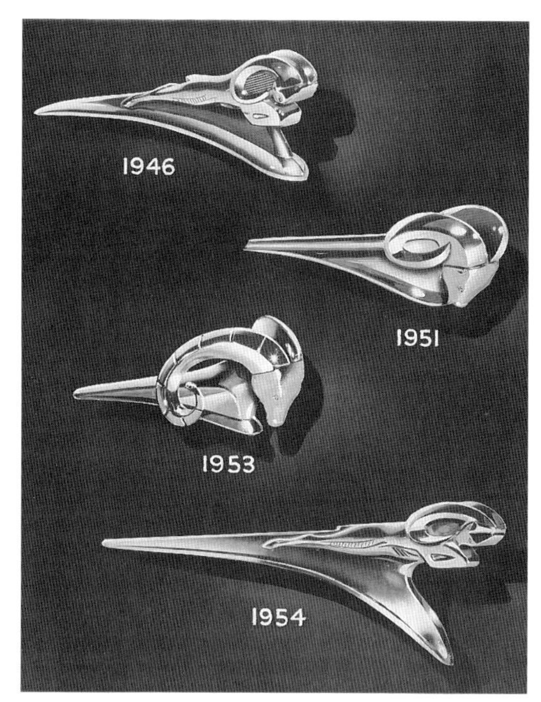

The Dodge Ram, designed in 1930 by Prof. Avard Fairbanks of the University of Michigan, first appeared on the 1932 models. Although the ram mascot ceased on cars after 1954, the symbol remains today in one form or another on most Dodge products.

Production: model year figures for the various models, submodels and body styles under discussion.

Specifications: basic engine, chassis, and drivetrain figures, along with wheelbases, weights and tire sizes, and typical performance figures.

Investment Ratings

I so much enjoyed Graham Arnold's *Illustrated Lotus Buyer's Guide* that I determined to adopt my investment ratings from his. Arnold made his book entertaining enough to be enjoyed by the nonspecialist as well as the enthusiast for one particular marque. I believe this is a good thing. Instead of the old one- to five-star general rating I and others have used in earlier Buyer's Guides, I have adopted Arnold's "fun, investment, and anguish" ratings on a scale of one to ten.

It occurred to me that the star system was fairly imprecise, not broadly indicative of a car's character. A Dodge Hemi Charger is great fun and a prime investment, for example, and would get five stars in the old system, but tune-ups will be frequent and any major engine work will be tremendously costly. A star rating would not reveal this, but a "9" in the "anguish" column is more suggestive.

Dodge Omnis also exemplify the advantages of the Arnold rating system. The best of the standard breed rate 5 (fun), 3 (investment), and 6 (anguish), the "6" being for the price of replacement parts and the likelihood of a complicated specification to go wrong as time passes. But the Shelby version rates 7/6/6.

Graham Arnold's anguish factor was defined as reflecting two aspects of ownership: "first, the likelihood of outright failure and frequency of nagging faults; and second, the shortage of spare parts and/or information, even sympathy." I have adopted his definition wholesale.

This rating system remains imprecise, as all rating systems are; but it's far more relevant than the star system and I hope you like it.

Price History

A stockbroker can go to jail by claiming specific future performance for a security, but there's nothing wrong with examining where that security has been in the past. Thus I present here the price history of top show-quality cars over the past dozen years, specifically the approximate high price of each model in 1982, 1988, and 1995, as a means of gauging its likely future behavior on the old car market.

My baseline year is 1982 because that's as far back as I can find reliable asking prices, auction and sales information. The midpoint, 1988, also happens to be the high point of the car-investment mania that has since, thankfully, leveled off. Along with the 1995 typical price for a 95+ point example, I have calculated the thirteen-year rate of return, assuming you invested the 1982 price then and sold for the 1995 price today.

Thus, for example, if you bought a 1968 Coronet R/T in prime condition for, say $12,500 in 1982, and your car is worth $27,500 today, that represents a compound rate of return of 6.8 percent on your money, which is better than most banks are paying. (In fact, you'll find the Coronet R/T and most other performance Dodges of its era returning much better than that on your money, if you bought them right, and far enough back.)

What is apparent from these calculations is how little Dodge as a marque was affected by the old car investment hoopla of the late 1980s and subsequent market reversals of the early 1990s. True, the odd Hemi was bid way up on occasion, but virtually everything else in the Dodge field was untouched. While makes like Ferrari, Aston Martin, and Mercedes-Benz took dives during the inevitable market downturn of the 1990s, most Dodges just kept plodding along, gaining in value. You will be surprised at the return on investment provided by almost every Dodge from the mid-1950s to the mid-1970s; it sure beats most mutual funds, including mine.

Cars, of course, are not CDs or stock portfolios; they require maintenance, insurance, parts, and service. Also, the price figures are highly arbitrary, taken from three or four sources and averaged. Prices apply only to very fine, condition 1 original or (more likely) restored cars, which always command far more than the same models in mediocre condition.

It is not possible individually to measure the overhead involved in twelve years' ownership of any specific car. Obviously, running costs are going to be higher for a Charger Daytona Hemi than they are for a '50 Diplomat. On the other hand, the rate of return does not take into consideration the intangible fun of ownership, which, after all, counts for something. And maybe the two balance each other out.

Richard M. Langworth
Hopkinton, New Hampshire
February 1995

Chapter 1

Fun	Investment	Anguish
5	3	6

Fours
1914–28

History

The Dodge brothers, John and Horace, built their first 247 cars for sale to the public in 1914, John suggesting that the market would take many more: "Just think of all the Ford owners who will someday want an automobile."

Machinists born and bred, John and Horace had taken an early and profitable interest in Ford, but had become disenchanted. Owning a one-tenth interest in Ford Motor Co., they had redesigned and improved major components, including rear axles and engines, beginning in 1902. Dodge Brothers also built components for other makers, and by 1910 their factory in Hamtramck, Michigan, was the largest automotive parts company in the world. Henry Ford's determination to vastly expand his River Rouge plant in an effort to bring much more component manufacture under his own roof, determined the Dodges to break their relationship. By mid-1914 they had organized their own car company. So well respected were John and Horace that they received 22,000 applications for dealerships before the first Dodge was built. If Packard had won hearts with their slogan, "Ask the Man Who Owns One," Dodge had the rejoinder: "It Speaks For Itself."

Dodge Brothers built nothing save four-cylinder models. In the words of the old Lucky Strike commercial, their cars had no fads, frills or fancy stuff; but, for about $200 more than the despised Model T, they offered granitic durability through the use of solid components: the Dodge's strong cast iron L-head four; an all-steel welded open Budd body (the first such to be mass-produced); three-speed selective transmission; a twelve-volt electrical system (also novel); and a 110in wheelbase. The wheelbase went to 114 in 1917 and 116 in 1924 which, wrote Beverly Rae Kimes, was "just about the only change made to the Dodge Brothers for the next decade." This was not exactly accurate—the cars acquired speedlines, hood louvers, more prominent radiators, and numerous mechanical improvements in the 1920s—but close enough. A Dodge Brothers looked very much up to date in 1914, but very much the dowager by 1925, when the firm was sold to the New York banking conglomerate Dillon, Read and Co. for $146 million.

John and Horace Dodge did not live to retire millionaires: both died in 1920, Horace from drink. Dodge Brothers was run on behalf of their widows by Frederick Haynes until the Dillon sale. Three years later, on July 31, 1928, Dillon sold Dodge to its present owner, Chrysler Corporation, for stocks and cash totaling $170 million—a cool $4 billion in today's money. But it was worth the price.

While Dodge never rivaled Ford for numbers, it was clearly a going concern. In their first full year of production Dodge Brothers ran third behind Ford's half a million cars

Horace (left) and John, the Dodge Brothers, driven by George Brown in the first Dodge, November 14, 1914. This photo was taken in front of the John Dodge estate, 33 East Boston Blvd., Detroit.

and Willys-Overland's 90,000. In 1920 with 141,000 they were second only to Ford, the best Dodge showing ever in the annual production race. Though routinely outproduced by the Ford, Chevrolet, Hudson, and Willys in the early twenties, Dodge remained profitable, and never finished below sixth through 1926.

Identification

The few cars built in 1914 were officially 1915 models: a touring was the sole body. A roadster was added in 1915, and a center-door sedan (none are known to exist) in 1916, which also saw the adoption of the multiple disc clutch. For 1917 the wheelbase went to 114in, although a few Dodges were built on the 110 chassis early that year; and a "Rex [Manufacturing Co.] convertible" (removable hardtop) was added for both roadster and touring. The '18s were continued unchanged. During the 1919 model year windshields became narrower and the Rex was joined by a five-window coupe, which replaced it entirely in 1920. During 1920, radical (for Dodge) slanted windshields were phased in and rear fenders were made longer on touring bodies.

In 1921 most softtop side and rear windows were oblong, instead of the pointed "cathedral style" used since 1914, and some '21s had a headlamp tie-bar. Also for 1921, closed cars had full-width seats, a heater was added, and the wheel size was reduced. Open models were unchanged in 1922 but sedans had lower roofs and straighter beltlines, and steel discs replaced the wire wheel option (artillery spokes were still normal equipment). All-steel closed bodies were an industry first in 1923 and custom bodies were offered on Dodge chassis for the first time: sedans, coupes, and a depot hack.

Nineteen twenty-four was a year of major change, with higher radiators and drum-type headlamps, louvered hoods and a standard-equipment rear brake light. Equipment options this year included nickel-plated radiator and bumpers, windshield wipers, motometer radiator caps, and bright metal runningboard step plates. Custom-built by Dodge were taxis and a landau sedan. Appearance was unchanged in 1925 but automatic windshield wipers, opening rear windows, cowl vents, and a one-piece windshield (on all but the business sedan and standard roadster) were new. In 1926 the "H" pattern three-speed shifter replaced the previous fore-aft shift pattern and triple door hinges were adopted. The '27s were much the same but offered a speedometer, ammeter, electric horn, headlamp dimmer, and inside rearview mirror as standard equipment.

The last Dodge four for many years was 1928's "Fast Four," Series 128 and 129 in standard and deluxe models, equipped with full instruments, a Stewart carburetor, and North-East ignition. The 129 had four-wheel "steeldraulic" brakes.

Appraisal

Dodge Brothers cars can be summarized in a few words: dependable (the adjective was early used in advertisements), uninspiring to look at, mechanically straightforward, built with integrity, and made to last. If their performance and looks were mediocre, their prospects for a long life were undoubted. The problems that beset owners of these Dodges today are those familiar to any owner of a genuine antique car (referring to that period through 1930): the difficulty of authenticating correct equipment and parts, obtaining them, and the cost of reproducing what cannot be found. In satisfactory running order, Dodge Brothers provide motoring enjoyment for a relatively modest investment.

Values and Prospects

Antique cars in general have not appreciated much on the old car market, and Dodge Brothers are no exception. The typical closed model (if you can find one—they're not common) should not set you back more than $5,000 in showworthy condition; nice open cars cost about double that, and special open bodies of the mid- to late-1920s can occasionally run up to $20,000. But surely that's the limit, and likely to remain so for the foreseeable future. The models to look for are 1923–28 customs, the 1926 sports roadster (dark green body, natural finish wood spoke wheels, bumpers front and rear, nickel radiator, gray leather, bullet headlamps, and trunk compartment), the 1927 convertible cabriolet (dark green body, landau irons, leather upholstery, door glass winders, rumble seat), and the 1928 Fast Four cabriolet. These are worth at least 20 percent more than the comparable body in standard and earlier models. Appreciation rates should match inflation throughout the next decade.

Price History

95+ point condition 1	1982	1988	1994	Return
1917 touring	$7,500	9,000	14,500	5.6%
1920 sedan	5,000	5,500	6,000	1.5%
1927 sport roadster	10,000	12,500	15,000	3.4%

Production

1914	1915	1916	1917	1918	1919	1920
247	45,000	71,400	90,000	62,000	106,000	141,000
1921	**1922**	**1923**	**1924**	**1925**	**1926**	**1927**
80,000	142,000	151,000	193,861	201,000	265,000	146,000
1928						
about 20,000						

Specifications

Engine

Type: cast iron L-head four
Bore & stroke (in): 3 7/8 x 4 1/2
Displacement (ci.): 212.3
Brake horsepower: 35 (NACC hp 24), 44in 1928 (NACC hp 24)

Chassis and Drivetrain

Transmission: selective sliding three-speed
Suspension: solid axles front and rear, long tudinal leaf springs

Measurements

Wheelbase (in): 110 through 1916, 114 in 1917–23, 116 in 1924–28
Curb Weight (lbs): 2,200 (open models), 2,500–2,800 (closed models)
Tire Size: 4.00x33 through 1921; 4.00x32 on 1922–23 and 1924 standards; 5.77x30 from 1924 Specials through early 1926; 5.77x31in late 1926; 5.25x31 in 1927; 6.00x31 on 1928 Series 128; 5.00x29 on 1928 Series 129.

Performance

Acceleration, 0–60mph: NA
Top speed (mph, approx.): 50–60
Fuel mileage (range): 14–18

Chapter 2

Fun	Investment	Anguish
5	2	5

Sixes

1928–34

History

The hard-drinking Dodge Brothers found their logical successors in the hard-driving Chrysler team. Stories abound of the Chrysler takeover of Dodge in July 1928, the best one being about how Chrysler lieutenant K.T. Keller changed the signs overnight to read "Chrysler Corporation—Dodge Division." Another tale recalls Clarence Dillon, phoning Chrysler the next morning to say Dodge was in good order and could "run itself for three months," and Walter Chrysler's reply: "Hell,

Classic lines and a handsome twin-bar front bumper distinguish this handsomely painted 1928 Senior Six sport cabriolet, a model that appeared in the spring of that year. The "sport" package included six wire wheels (two sidemounted spares), oversize headlamps on a nickel-plated light bar, nickel cowl lamps, and a folding trunk rack.

Clarence, our boys moved in yesterday." (But the "Dodge Brothers" badge didn't entirely disappear until 1939.)

In the business world it was like General Motors being bought by Studebaker. The Wall Street consensus, wrote Thomas MacPherson in *The Dodge Story*, was that Chrysler had bitten off more than he could chew, and people began "making jokes about Chrysler's folly." But Walter Chrysler replied, "Buying Dodge was one of the soundest acts of my life . . . without Dodge there would be no Plymouth." (The Hamtramck plant increased Chrysler's floor space by 500 percent and gave it world class foundry and forge facilities needed to build Plymouths in quantities to present a serious challenge to Ford and Chevrolet.) Without Dodge there would still have been a DeSoto, though—some believe Chrysler planned the DeSoto to scare Dillon into selling him Dodge, though by the Dodge sale DeSotos were already in production. But Dodge would outlast DeSoto in the end, and contribute mightily to making Chrysler one of the "Big Three." In the forties and early fifties, Chrysler actually ran ahead of Ford in production.

A six-cylinder Dodge Brothers had already been planned under Dillon's management when Chrysler took over; it dominated sales in 1928, outsold the shortlived Dodge straight eight during 1930-33, and established itself as Dodge's only motive power from 1934 until 1953. Substantially faster and better looking than the previous fours, it also cost more, and thus produced lower sales volume, particularly after the Wall Street crash in 1929. But as the market began to revive after 1932, the six proved a canny choice. A growing number of buyers were able to afford more than a Ford-Chevy-Plymouth, and by 1933 Dodge was firmly in command of fourth place in production behind the "low-price three."

These Dodges took two essential forms: upright and foursquare through 1932; more rakish starting in 1933, with slanted radiators and longer, flowing fenders. In 1928 there were three lines: Standard, Victory and Senior Six on 110in, 112in, and 120in wheelbases. This made the Senior Six substantially more Dodge than the other two models, which is reflected in its value among collectors today. The 110 chassis was dropped at the beginning of 1929, but a year later with the deepening depression a 109in second series "New Six" was announced with prices starting at $855, and this helped preserve Dodge's market share

Only 8,523 of the 1933 Series DP two-door sedan were built, and it is much more desirable than the more common four-door sedan. Avard Fairbanks's Dodge ram mascot was in its second year.

through the worst of the economic doldrums. The famous Dodge Ram mascot, designed by Avard Fairbanks, was introduced on the 1932 Dodge, and has been around in one form or another ever since.

Neoclassic styling began to creep into the Dodge formula with the 1933s, whose vee'd radiator, nicely angled and curved toward the bottom, was similar to Packard's Light Eight of the year before. This was also the year Dodge aligned its new models with the model year—in 1932 there had been eight different overlapping lines. The '34 was swoopier looking, with skirted fenders, horizontal hood vents, chrome bullet headlamp housings and longer wheelbases for both the standard and deluxe lines. Most Dodges were on a 117in wheelbase this year, but two models, the Series DS Aero Brougham and convertible sedan, look rakish on a 4in-longer chassis with their "slantback" styling.

Identification

The 1928 Standard, Victory, and Senior Six can be identified by their wheelbases: 110in, 112in, and 120in, respectively. The Standard and Victory Sixes were carried over with little change for the first part of 1929, while the Senior Six was given vertical instead of horizontal hood louvers and a new silver-finish instrument panel. Dodge was introducing new models in midcalendar year and switched to series designations in 1929 with the DA Six, a facelifted Victory Six with chrome instead of nickel plating, headlamps mounted on the tiebar, and a longer hood with a narrow bright molding on its rear edge.

From July 1929 the Series DA was joined by the DB, an extension of the Senior Six, but with four-speed transmission and rubber engine mounts. These cars can be identified by their vertical chrome headlamp supports. The DD six (January 1930; the DC was an eight) had a shorter wheelbase, wider radiator, and "VV" or "Vision and Ventilating" windshield.

Late in 1930 came the DH, on a longer wheelbase, with a wider and deeper shutter-type radiator, twin cowl vents, a double front bumper with V-shaped upper bar, and ebony-finish instrument panel. The DH and DD were continued for 1931.

Dodge sixes for 1932 comprised the previous DD and DH Series and the new DL (sloped windshield, ram mascot, curved-V double bar bumper, chrome headlamps, longer and lower body lines). For 1933, model introductions lined up with the model year and the sold six-cylinder Dodge was the Series DP, with its curved and canted V-type radiator, suicide doors, chrome headlamps, and single bar bumper. The DP came with two wheelbases, 111 and 115.

Dodge having dropped its eights for 1934, the line of sixes was again expanded, with three series: the restyled DR, with skirted fenders and horizontal hood louvers; the DS, a long-wheelbase offshoot comprising Aero Brougham and convertible sedan models; and the DRXX, a stripped version of the DR. Cost-cutting features of the DRXX are immediately obvious: body colored radiator shells, no pinstriping, no safety glass.

Appraisal

The early Dodge six was a good looking, reasonably decent performer that offers many pleasures to the aspiring owner. They don't cost an arm and a leg to buy, they're simple and straightforward to tinker with, and parts are not entirely impossible even though it is nearly seventy years since the first one hit the streets. They are recommended to car lovers who like the products of this period, but not investor-types.

Models to look for are the 1933–34s in

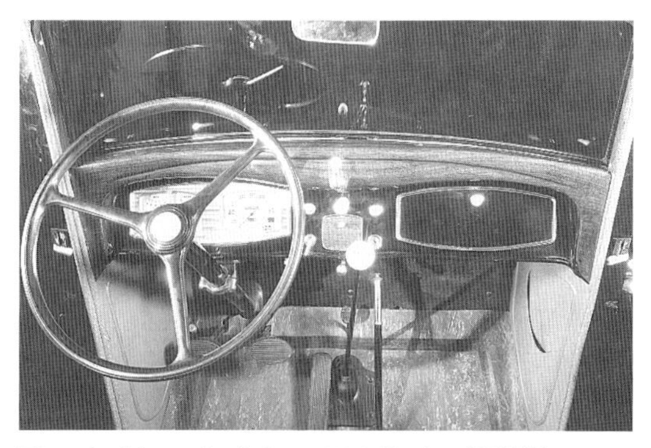

The dashboard of the 1934 Series DRXX was a fresh design: a good-looking panel set into a real tree-wood molding. Black-on-white instruments were a long-running Dodge tradition. This was the last year for "suicide doors."

A highly desirable thirties Dodge is the 1934 Series DS Special convertible sedan on the 121in wheelbase. Only 350 of these beauties were built, and survivors are rare. Art deco influence was already evident in the headlamp pods, hood louvers, and smoothly curved rear quarters.

An unidentified 1934 artist's drawing of a rumble seat coupe illustrates the more modern look achieved with new style wheels and tapered headlamps. Among closed models, coupes are far better investments than sedans.

preference to the earlier models; convertible coupes and roadsters in preference to large open bodies like convertible sedans; the rare five-passenger 1932 Series DH sport phaeton (only 164 built); the equally rare 1934 Series DRXX 2–4-passenger coupe (a mere 105); and the 1934 Series DR seven-passenger long sedan (710). Open models in the early years generally saw fewer than 1,000 copies of each and are of course highly desirable, but don't pay overblown prices for these as they've dropped drastically in value over the past six to eight years. Four-door sedans had the lion's share of production by far in all years and are the least desirable body style; the Special or Aero broughams had limited production and are the best four-door buys.

Values and Prospects

The following price histories tell the stories of these Dodges. No Dodge is rated a "Classic" by the Classic Car Club of America (CCCA); but the flashy open models were subjected to the same kind of galloping prices as true Classics during the great old car bull market of the late 1980s, and several of them have come crashing down. The 1929 Senior Six, for example, virtually doubled in value in six years between 1982 and 1988, but at this writing it has depreciated by over 20 percent from its high. Interestingly, the early 1930s models have held up better, and the body style that seems to do best (as ever!) is the rumble seat convertible, the archetypal open car of the 'tween wars period. By comparison, the convertible sedan, a larger, clunkier body preferred by a smaller body of enthusiasts, has leveled off to barely its 1982 level. Cranking in the inflation factor, a DS convertible sedan has actually depreciated.

What does this mean in real life? Nothing, if making a big profit isn't your objective; indeed these handsome open Dodges are arguably more affordable by more people today than anytime in the last fifteen years. But speculators are off them, and they should not be looked upon as a high-yield security. Closed models, regardless of body style, seem stuck around $8,000, which in 1982 dollars is really only $5,000 or so: a good buy.

Price History

95+ point condition 1	1982	1988	1995	return
1928–34 closed body styles	$7,000	7,500	8,000	1.1%
1929 Senior Six spt roadster	14,500	26,000	20,000	2.7%
1930 DD phaeton	19,000	29,000	28,000	3.3%
1932 DL rumble seat conv.	19,500	27,000	26,000	2.4%
1933 DP rumble seat conv.	17,000	30,000	26,000	3.6%
1934 DS convertible sedan	20,000	31,000	21,000	0.4%

Sixes at a GlanceF

Model (intro.)	Model Years	Wheelbase	First year prices
Standard Six	1928–29	110in	$875–970
Victory Six	1928–29	112in	$1,030–1,170
Senior Six	1928–29	120in	$1,495–1,800
DA (11/28)	1930	112in	$925–1,065
DB (7/29)	1930	120in	$1,545–1,615
DD (1/30)	1930–32	109in	$835–935
DH (11/30)	1931–32	114in	$815–845
DL (11/31)	1932	114in	$795–945
DP (11/32)	1933	111in	$595–695
DP lwb (4/33)	1933	114in	$595–675
DR (11/33)	1934	117in	$665–765
DS (1/34)	1934	121in	$845–875
DRXX (6/34)	1934	117in	$645–745

Specifications

Engine

Type: cast iron L–head six
Bore & stroke (in): 3 7/8 x 3 7/8
Displacement (ci): 208
Brake horsepower: 58

Chassis and Drivetrain

Transmission: selective sliding three–speed, column shift
Suspension: ladder chassis, beam axles and leaf springs front & rear.

Measurements

Wheelbase (in): 109–121in; see following table
Curb Weight (lbs): 109–110in wheelbase cars 2,400–2,700; 111–114in wheelbase cars 2,500–2,900; 120–121in wheelbase cars 2,900–3,700
Tire Size: Standard and Victory Six 5.00x29; Senior Six 6.00x31; Series DA 5.00x29; Series DB 6.00x19; Series DD and DH 5.00x19; Series DL 5.50x18; Series DP 6.00x16; Series DR, DS and DRXX 6.25x16

Performance

Acceleration, 0–60mph: NA
Top speed (mph, approx.): 70–80
Fuel mileage (range): 12–17

Production

1928 (estimate): 45,000
1929: 121,457
1930 (estimates): 14,000 DA, 7,000 DB
1930–32 DD: 33,432 4dr sedans; 3,877 business coupes; 3,363 coupes; 772 roadsters, 620 convertible coupes, 542 phaetons, 899 chassis. Model year production about 20,000 for 1930, 13,000 for 1931, 1,000 for 1932.
1931–32 DH: 33,090 4dr sedans; 4,187 coupes; 3,178 business coupes; 164 phaetons (1932 only); 160 roadsters; 47 chassis. Model year production about 20,500 for 1931 and 20,300 for 1932.
1932 DL: 16,901 4dr sedans; 1,963 business coupes; 1,815 coupes; 224 convertible coupes; one five-passenger coupe; 126 chassis.
1933 DP: 69,074 4dr sedans; 11,236 business coupes; 8,875 coupes; 8,523 2dr sedans; 4,200 salon broughams; 1,563 convertible coupes; 980 chassis.
1934 DR: 53,479 4dr sedans; 8,723 business coupes; 7,308 2dr sedans; 5,323 coupes; 1,239 convertible coupes; 710 seven-passenger sedans; 1,475 chassis.
1934 DS: 1,397 Special broughams; 350 convertible sedans; 3 chassis. 1934 DRXX: 9,481 4dr sedans; 3,133 2dr sedans; 2,284 business coupes; 104 coupes.

Chapter 3

Fun	Investment	Anguish
7	7	6

Eights
1930–33

History

A lot of redundant new models came out of Detroit in the wake of the 1929 Wall Street stock market crash and subsequent depression, most of them because they'd been planned when business was good. Cadillac probably wouldn't have built twelve- and sixteen-cylinder cars had it known what was coming; Studebaker wouldn't have bought Pierce-Arrow; Oldsmobile would have forgotten the Viking; Packard would have built the medium-priced One Twenty sooner—and Dodge would almost certainly have scrubbed plans for a straight eight.

The eight was born out of Walter Chrysler's initial impulse to position Dodge Division between DeSoto and Chrysler. (It wasn't until the mid-1930s that Dodge was slotted between Plymouth and DeSoto.) In the early 1930s most DeSotos sold for well under $1,000, and while there were Dodges in that price bracket there were also Dodges well above it: the Senior Six, for example, cost upwards of $1,600. The advent of hard times reduced demand for an up-market Dodge eight dramatically. After selling close to 20,000 for the 1930 model year dealers found eights almost impossible to move, and production fell off to about 4,000 the following year. By 1933, when the eight was phased out, volume had skidded to 1,600.

The Series DC, with its five-main-bearing 220in "stroker" eight developing 75hp, was billed as "more attractive, more powerful, longer, and roomier" (than the Dodge six) upon its debut in January 1930. It looked impressive on its 114in wheelbase, and Dodge made it instantly distinguishable from the Six with a chrome "Dodge 8" badge, wide shell radiator, and thick hood louvers. A raft of standard equipment included Delco-Remy ignition, hydraulic brakes, a slanting windshield with visor, a redesigned instrument panel, hydraulic shocks, and rubber shackled springs. Aside from the *de rigueur* coupe and sedan there were three open body styles, roadster, phaeton, and convertible coupe; prices started at $1,095 for the roadster, which was certainly value for money. The problem was that nobody *had* any money. Open models saw very low production.

Prices were cut by as much as $150 in mid-1930, but the bottom had fallen out of the market and the Series DC was carried into 1931 as a rather forlorn hope. A new instrument panel with three control buttons (July 1931) was among the few obvious changes. The "true" 1931 model was the Series DG, introduced early that year, with two open and two closed bodies on a longer wheelbase. It accounted for about 11,000 units through the end of the year. In January 1932 came the yet longer DK or "New Eight," with a larger frame and longer wheelbase, sloped windshield, interior sun visors, and wire wheels. Comprehensively equipped with "Floating

Power" motor mounts, silent second gear transmission, freewheeling and automatic clutch, it was the best looking eight yet.

For 1933, Dodge fielded the Series DO on the DK's wheelbase, but with all new styling tricks which improved its appearance: single bar bumpers, vertical hood louvers, chrome-plated bullet headlamps, sloped windshield, and "beaver tail" rear. The DO came in two closed and five open body styles, but by now demand was nonexistent and it proved of the rarest Dodge models ever built: only 1,652, most of them sedans.

Identification

The Series DC carries a "Dodge 8" badge in the upper center of the radiator; six-wheel equipment (wire wheels) was optional, and standard on the phaeton model. The DC also has obviously fatter hood louvers than on six-cylinder Dodges. From July 1930, DCs were fitted with twin cowl vents, oblong windows and a new radiator emblem with a figure 8.

The 1931 Series DG had lever-operated radiator shutters, swinging windshield, a 118.5in wheelbase, and was the first eight to use the Dodge ram mascot. Its wheelbase was 4in longer than that of the DC. Both the DC and DG were carried on into the 1932 model year.

Announcement of the Series DK "New Eight" in November 1931 gave Dodge no fewer than three different eight-cylinder '32 models, which was three too many; but the DK was a lovely automobile. Mounted on a stretched, 122in chassis with longer, lower lines, it looked and was the queen of the Dodge line. Six models were offered, two open, but unfortunately demand was negligible and barely 6,000 were produced. When Dodge lined up models with model years for 1933, its lineal successor was the Series DO, the final Dodge eight until after World War II, easily recognizable from its curved radiator and beaver-tail rear. This eight also featured triple-beam headlamps.

Appraisal

By Dodge standards these are big, strong, powerful cars, and in terms of performance they're the best thing Dodge offered until 1953—forebears, if you will, to the great Dodge road cars of the 1950s and 1960s. Fuel

Probably the nicest Dodge Eight was the 1931-32 Series DG, with lever-operated radiator shutters, swinging windshield, a lengthy 118.5in wheelbase. The '32, shown here, was the first eight to use the Dodge ram mascot. The rumble seat roadster sold for $1,095; fewer than 3,000 were built.

mileage is not their forte, but if smoothness of operation and keeping up with modern traffic is important, you'll like the eight. It was also the most luxurious Dodge for some time, and very fully equipped. Styling is excellent, beautifully proportioned. Though not "Classics" by the CCCA definition, eights are serious collector cars. Parts availability is only fair and engine parts are scarce.

Values and Prospects

Open models of the Dodge eight were not subject to the market upheavals that affected the sixes mentioned in the previous chapter. With only one minor exception, all the sample models I tracked through twelve years of price guides are worth more today than they were seven years ago at the height of the Reagan recovery. None have produced significant rates of return on investment, but all have held or increased their value at a time when former high flyers and auction stars like Ferrari, Duesenberg, and Cadillac were tumbling from the heights. It seems to me that you can look at Dodge eights from the same financial angle as you view the typical CCCA Classic: not leaping in value, but trending steadily upward like a good blue chip stock. Almost certainly, the eight you buy today will be worth a fair amount more five or ten years from now, even if it's a closed model.

Dodge eights to look for are the coupes and virtually any open model you can find. Production of every body style save four-door sedans was extremely low. The 1932–33 Series DK and DO are the scarcest and the 1932 DG phaeton is particularly desirable—the last of that body style built by Dodge. Some of the production figures will get collector juices running (see below), especially for convertible coupes and convertible sedans. The problem is that there aren't many on the market. Supply is well behind demand.

Price History

95+ point condition 1	1982	1988	1995	return
1930–33 closed bodies	$8,500	9,500	10,500	1.8%
1930 DC roadster	20,000	29,000	28,000	2.8%
1931 DG phaeton	26,000	29,000	31,000	1.5%
1932 DK conv. sedan	23,000	28,000	30,000	2.2%
1933 DO rumble seat conv.	22,000	25,000	33,000	3.4%

Eights at a Glance

Model (intro.)	Model Years	Wheelbase	First year prices
DC (1/30)	1930–32	114	$1,095–1,225
DG (1/31)	1931–32	118.5	$1,095–1,170
DK (1/32)	1932	122	$1,115–1,395
DO (1/33)	1933	122	$1,115–1,395

Specifications

Engine

Type: cast iron L-head eight
Bore & stroke (in): 2.88 x 4.25
Displacement (ci): 220.7
Brake horsepower: 75

Chassis and Drivetrain

Transmission: selective sliding three-speed, column-shift
Suspension: ladder chassis, beam axles, and leaf springs front & rear.

Measurements

Wheelbase (in): 114–122in; see following table
Curb Weight (lbs): DC 2,800–3,000; DG 2,900–3,300; DK 3,500–3,700; DO 3,500–3,900
Tire size: Series DC, DG and DL 5.50x18; Series DO 6.50x17

Performance

Acceleration, 0–60mph: NA
Top speed (mph, approx.): 85–90
Fuel mileage (range): 10–15

Production

1930–32 DC: 20,314 4dr sedans; 2,999 coupes; 728 convertible coupes; 598 roadsters; 234 phaetons; 123 business coupes; 253 chassis. Model year production about 20,000 for 1930; 4,000 for 1931 and 631 for 1932.
1931–32 DG: 8,937 4dr sedans; 2,936 roadsters; 2,181 coupes; 500 convertible coupes; 20 chassis. Model year production: 9,520 for 1931, 2,344 for 1932.
1932 DL: 16,901 4dr sedans; 1,963 business coupes; 1,815 coupes; 224 convertible coupes; 12 convertible sedans; one five-passenger coupe; 126 chassis.
1933 DO: 1,173 4dr sedans; 212 two/four-pass. coupes; 159 five-pass. coupes; 56 convertible coupes, 39 convertible sedans; 13 chassis.

Chapter 4

Fun	Investment	Anguish
5	3	4

Sixes

1935–42

History

Dodge Division was spared a catastrophe in 1934: it didn't produce an Airflow. Chrysler and DeSoto did, and lived (just barely) to regret it. Dodge cars were stolid and conservative, adopting a popular potato shape in 1935, and the Hamtramck Division hung on gamely in the production race, building a record 295,000 cars for the recovery model year 1937. Recession year 1938 sent sales tumbling again, but Dodge was never lower than seventh place and would have gone over 300,000 for 1942 had World War II not closed down production in February.

An important change which benefited Dodge in this period was its change in market position from both sides of DeSoto to a distinctly cheaper make from 1934 onward. Having opted for a single six-cylinder engine (218ci, not changed until it was stroked to

The 1937 Westchester Suburban was the first woody-wagon offered as a passenger car rather than as a commercial vehicle. Bodies came from U.S. Body & Forging Co. in Buffalo, New York. The rounded front end styling was the work of Ray Dietrich. Production of this model was a mere 375.

Smooth, rounded lines distinguished the 1935 Dodges from their upright predecessors. This seven-passenger sedan was one of the low-volume models (barely 1,000) and would be a nice find today. A luxury interior with jump seats and reading lights was fitted.

230ci for 1942), the Division produced a host of imaginatively named series: "New Standard" for 1934, "New Value" for 1935, "Air-Styled Beauty Winner" for 1936, "Luxury Liner" for 1939. Chassis were usually around 115–117in in wheelbase, with a handful of long-wheelbase models still offered, including the first Dodge Caravan, a 128in wheelbase five-passenger sedan (1935), a long-door job with a built-in trunk and curious abbreviated rear quarter windows.

Mechanically Dodge was evolving fast in this period. Overdrive, steel artillery wheels, independent front suspension ("Floating Cushion Wheels") and "Draft-Free" ventilation were features for 1934. The following year saw a kind of precursor to "Cab-Forward Design," with the engine shifted 8in forward and the front seat 6in forward, creating a capacious rear passenger compartment with a seat well ahead of the rear axle. A retrograde step was the reversion to beam front axle and leaf springs in 1935, a setup which lasted through 1938. Nineteen thirty-six was an evolutionary year in styling and mechanical details, although the all-steel top wired for a radio antenna was an innovation.

Dodge's banner sales year of 1937 was made possible in part by many popular mechanical innovations, including low driveshaft tunnels, built-in defroster vents, and several features emphasizing Dodge's early interest in safety: no-snag door handles, recessed dash knobs, flush-mounted instruments. Most important, though, was the first use of insulated rubber body mounts. The mildly altered '38s were the last Dodges to use the "Dodge Brothers" badge with its superimposed double triangles—and the first to feature Dodge's famous Fluid Drive. This was not a common option in prewar years and will be described in Chapter 5.

Dodge's Silver Anniversary and the World's Fair in 1939 were celebrated by a two model line-up for the first time in five years and fresh styling by Raymond H. Dietrich, a former coachbuilder who brought the cars a fresh, art deco look with flush fitting headlamps, fastback styling, integrated trunk, two-piece vee'd windshield and a divided horizontal grille that wrapped around the inner front fenders.

Though there were no open or long-

The 1937 four-door sedan (as distinguished from the touring sedan, which had a bustle trunk) was a graceful Dodge with swoopy fastback rear end styling. About 7,500 were built and sold at $820.

wheelbase bodies in 1939, they returned the following year, along with cleaner styling, more chrome, and a wheelbase stretched to nearly 120in. Dodge had now acquired the essential body that it would take into the postwar years. The 1941s were given a larger grille stretching from headlamp to headlamp and Deluxe models with electric windshield wipers and Airfoam seat cushions. A significant facelift was applied in 1942, with an egg-crate style center grille and elongated fenders, and an extremely handsome symmetrical, woodgrained dashboard.

Identification

The '34s continued 1933 styling but with a vee'd radiator, skirted fenders and large bullet-shaped headlamp shells and a "dip" in the center of the front bumper. The '35s were completely restyled and much rounder, with a narrower, more raked grille and painted headlamp shells. Using the same basic body, the '36 featured a rounded grille, twin rows of transverse hood louvers and built-in horn housings below the headlamp shells. A divided grille with vertical center moldings marked the 1937 models; the '38s were almost the same but can be distinguished by straight horizontal hood louvers which match the pattern of the grille.

Art deco styling—a divided, horizontal bar grille following front end contours and flush-fitting headlamps—arrived for 1939; the horizontal theme was also applied above the grille at the leading edges of the hood. For 1940 the front end was simplified, with a one-piece, thin horizontal bar grille divided by a thicker horizontal bar. The divided grille reappeared in 1941 but extended to the headlamps on each side and the windshield and backlight were larger. For 1942, the grille was a single, headlamp to headlamp assembly, horizontal except for an egg crate section in the center.

Appraisal

Models to look for include convertible coupes in all years and the following unique but mostly-rare models: the 1934 convertible sedan on the longer, 121in wheelbase was the last of this body style on a really long wheelbase, although the convertible sedans were also built for 1936–37. In 1935 all the extended chassis went to 128in but contained only closed models, although the Caravan and seven-passenger sedan with sidemounts (production just 193 and 1,018, respectively) are desirable. From 1936 to 1938, the long wheelbase was confined to seven-passenger sedans and limousines only; the limos are by far the

Here's a slick prewar Dodge with a highly desirable body, the Hayes-built 1939 town coupe. Window treatment is reminiscent of the famous Cadillac Sixty Specials. Hayes built 1,000 of these beautiful bodies for Dodge, and similar ones for Chrysler and DeSoto. Original list was $1,055.

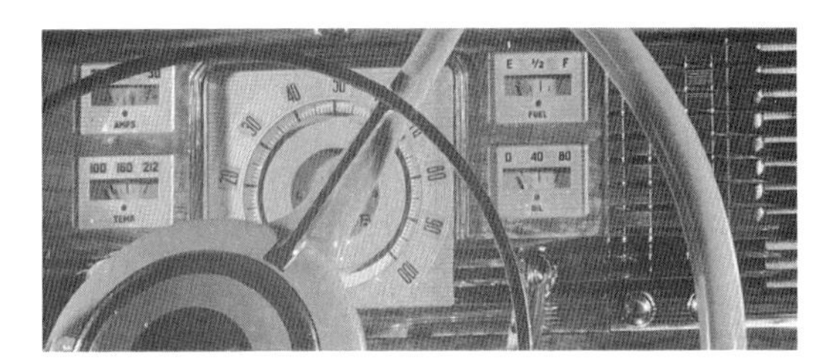

The Dodge instrument panel was under continual revision in the last years before the war, as these photographs of the 1940 and 1941 versions indicate. The '40 used lots of chrome and woodgrain; the "Safety Light" speedometer contained a ring that glowed different colors depending on speed. The '41 switched to white-on-black instruments and a more symmetrical pattern and an unconventional steering wheel/horn bar.

scarcer. Although no seven-passenger bodies were built in 1939, they returned in 1940 on the longest wheelbase of any Dodge in history, 139.5in; a handful were even built in 1942, and the model continued in the first few postwar years. Dodge also sold bare chassis to hearse, ambulance, and station wagon builders; these cars are very scarce and, of course, desirable.

The '39 Dodges seem to many collectors distinctly better looking than their recent predecessors, and when their handsome styling is combined with such models as fastback two- and four-door sedans or the Town Coupe (body by Hayes, probably only about 300 built), a very nice Dodge resulted.

Summary and Prospects

These Dodges have been good solid investments over the past decade. Being far outside the speculative area, their appeal is primarily to collectors, and they have been almost unaffected by the price fall-out that affected cars of greater pedigree and higher notoriety. Every one of them is worth a reasonable amount more now than it was worth in 1988, the year such auction stars as the '59 Cadillac Eldorado hit its peak (it has since de-

The '42 model you're most likely to encounter, the Custom four-door sedan, saw 22,055 copies which sold for $1,048 apiece. The chief distinction for this model year was the heavier, full-width grille treatment.

By contrast to the sedan, the '42 long-wheelbase models were extremely limited. Shown is a seven-passenger sedan (production totalled 201), which came with jump seats and sold for $1,395. The limousine, with leather front seat and division window, cost $80 more but saw only nine copies.

preciated by at least half its 1988 value). Late 1930s and early 1940s Dodges compare to the Caddy as the tortoise does to the hare.

Now that we've established that you're not throwing your money away, are there any distinctions to be made? Not many, but curiously enough the standard wheelbase closed body styles (coupes mainly) have appreciated faster in value than any of their other, rarer, ostensibly more desirable linemates, like limos and convertibles. But none of this means much to the prospective buyer of one of these cars: that buyer is in the game for the love of the vehicle, the era, the style, so what does it matter how much it'll gain by 2010? To meet the purposes of this section, however, let's just say that an investment in any 1934–42 Dodge is a sound decision you are unlikely to regret at the present state of the market. I would not, however, pay anything like $9,500 for a plug ordinary four-door sedan—a Hayes Town Coupe, maybe.

Price History

95+ point condition 1	1982	1988	1995	return
closed bodies, std. wheelbase	$5,000	6,000	9,500	5.5%
closed bodies, long wheelbase	7,500	8,000	9,500	2.0%
1934 convertible sedan	20,000	21,000	29,000	3.1%
1936 convertible coupe	17,000	24,000	27,000	3.9%
1938 convertible sedan	18,000	22,000	25,000	2.8%
1939 Town Coupe by Hayes	6,500	7,500	10,200	3.8%
1940–42 convertible coupe	15,000	22,500	25,000	4.3%

Specifications

Engine

Type: cast iron L-head six
Bore & stroke (in): 3.25 x 4.38 (4.63in 1942)
Displacement (ci): 217.8 (230.2 in 1942)
Brake horsepower: 87, 91 (1941), 105 (1942)

Chassis and Drivetrain

Transmission: selective sliding three-speed, column-shift
Suspension: i.f.s. with coil springs 1934 and 1939-on; beam axles with leaf springs front & rear in other years

Measurements

Wheelbase (in): 117, long wheelbase 121 (1934); 116, lwb 128 (1935–36);115, lwb 132 (1937–38); 117 only (1939); 119.2 standard from 1940 to 1942; lwb 139.5 (1940), 137.5 (1941–42) Curb Weight (lbs): From 1934–42, standard wheelbase models 2,800–3,250; seven passenger sedan 3,100–3,700
Tire size: 6.25x16 in 1934; 6.00x16 in 1935–36; 6.50x16 in 1937–42 standard models, 6.00x16 long-wheelbase models

Performance

Acceleration, 0–60mph: 25
Top speed (mph, approx.): 75–85
Fuel mileage (range): 12–18

Production

1934 DeLuxe: 53,479 4dr sedans; 8,723 business coupes; 7,308 2dr sedans; 5,323 coupes; 1,397 long wheelbase Special broughams; 1,239 convertible coupes; 710 7-pass. sedans; 350 long wheelbase convertible sedans; 1,478 chassis.

1934 New Standard: 9,481 4dr sedans; 3,133 2dr sedans; 2,284 business coupes; 105 coupes; one chassis.

	1935	1936	1937	1938
4dr sedan	74,203	174,334	185,483	73,417
4dr fastback	33,118	5,996	7,555	714
2dr sedan	18,069	37,468	44,750	17,282
2dr fastback	7,891	2,453	5,302	999
coupe, business	17,800	32,952	41,702	15,552
coupe, 2–4 pass.	4,999	4,317	3,500	950
conv. coupe	950	1,525	1,345	701
conv. sedan	0	750	473	132
7 pass. sedan (lwb)	1,018	1,942	2,207	1,953
7 pass. limo (lwb)	0	0	216	153
chassis	1,258	1,910	2,514	2,301
Caravan (lwb)	193			
wagon	374			

	1939	1940	1941	1942
Special, 1941–42 DeLuxe:				
4dr sedan	32,000	26,803	49,579	13,343
2dr sedan	26,700	27,700	34,566	9,767
bus coupe	12,300	12,001	22,318	5,257
club coupe	0	0	0	3,314
DeLuxe, 1941–42 Custom				
4dr sedan	5,545	84,976	72,067	22,055
4dr fastback	270	0	0	0
2dr sedan	1,585	19,838	20,146	4,685
2dr fastback	270	0	0	0
bus. coupe	630	12,750	0	0
coupe	0	8,028	18,024	4,659
conv. coupe	0	2,100	3,554	1,185
7 pass. sedan (lwb)	0	932	604	201
7 pass limo (lwb)	0	79	509	
chassis	0	298	20	0
Town Coupe (Hayes)	300 est.			
Town sedan	4,047			

Chapter 5

Fun	Investment	Anguish
5	4	4

Custom & Deluxe

1946–49

History

Dodge got off to a slow start after the car manufacturers were permitted to resume civilian vehicle production as World War II wound down, and only 420 units were built before the end of 1945. This quickly picked up, and Dodge was usually in fourth place behind the Chevy-Ford-Plymouth trio in these years. Wartime studies had produced many interesting postwar styling proposals, featuring smooth, wraparound grillework and bumpers, thin door pillars, integral fenders and acres of curved glass. But the 1940 body had been barely amortized and in common

A '42 body with a new grille, the '46 Dodge was designed to get the company up and running with a minimum of sheet metal change, and this body style continued into early 1949. The Custom convertible had a production run of 9,500 for the entire period, but few seem to be around nowadays.

with almost all manufacturers, Chrysler Corporation decided to produce mildly facelifted 1942 models for the first several years after the war. The facelift was done mainly by stylists A. B. Grisinger, John Chika, and Herb Weissinger. Allowed bolt-on alterations, they opted for a new grille with wide horizontal and vertical bars forming a pattern of rectangles. Parking lights were square, located at either side of the grille; the Dodge nameplate was mounted above it. Mechanical changes included relocating the starter, which went from a foot pedal floor control to a button on the dash. Front brakes were equipped with double wheel cylinders, an in-line fuel filter and full-flow oil filter became standard, and the 1942 dashboard was glitzed up with a bit more brightwork. Fluid Drive, first seen in 1938, became standard after 1946.

This good-looking postwar Dodge club coupe (with a prewar Chrysler convertible) is owned by Roger Mease of New Jersey.

Although Chrysler and DeSoto also had Fluid Drive, Dodge was the lowest priced American car with this semi-automatic transmission, and deserves a discussion here. Fluid Drive eliminated 95 percent of shifting by combining a conventional clutch with torque converter and electrical shifting circuits: "a full range," as one writer recorded, "of potential transmission trouble." Actually it was pretty reliable. The conventional flywheel was replaced with a fluid-coupling torque converter which performed the usual flywheel functions, storing energy, smoothing power impulses, and carrying the ring gear which

Although pictured in the 1946 catalogue, the long-wheelbase Custom town sedan did not enter pruduction until 1947, and fewer than 4,000 were built through early 1949. Two of these cars were built as limousines, with division windows and leather chauffeur's seat.

The best proportions on early postwar Dodges are probably those of the three-passenger business coupe, with an extended deck to house lots of salesman's samples. Sold only in the Deluxe trim version, it accounted for 27,600 sales during 1946-49 and sold for $1,229-1,605.

meshed with the starter pinion. Lacking a clutch plate contact, a clutch was mounted in tandem. The fluid coupling was a drum filled with low-viscosity mineral oil. As the engine ran, a set of vanes on the inner drum casing rotated, throwing oil outward onto a facing runner that had its own set of vanes. The oil turned the runner, allowing a smooth flow of power and avoiding metal-to-metal contact. There were only two gear positions: Low (first and second gear) and High (third and fourth). Low was used only for fast starts or towing. Normally the driver shifted into High and pressed on the accelerator, the transmission shifting from third to fourth with a "clump" at 14mph. The clutch pedal was still there, but used only to change between Low and High, or for back-up.

The body styles offered were typical of the period—not for Dodge wood-trimmed Aero-sedans or station wagons. The price-leader DeLuxe line offered two- and four-door sedans and a business coupe (the two-door sedan sold for the same $1,299 as the business coupe, and handily outsold the four-door sedan). The better trimmed Custom contained three four-door sedans, a club coupe and a convertible. The two standard-wheelbase sedans differed in their window arrangement: the standard and more popular model had a third side window behind the rear door, while a smaller number of "Town Sedans" had closed rear quarters and a vent wing fitted next to the rear door window. Dodge also built two seven-passenger limousines, apparently just an experiment, and a handful of chassis for the professional car trade.

From December 1948 through March 1949, Dodge extended this initial postwar line of cars for the 1949 model year, pending introduction of the completely restyled '49s in April. About 42,000 of these "First Series '49s" were built. Although they are rare today, they proved bad investments for those who bought them new, as they were considered '48 models when it came time to trade and were depreciated accordingly.

Identification

Aside from more detail to the hood ornament starting in early 1947, there is no way to tell the 1946–48 Dodges (and the small extended run in early 1949) apart, except from their serial numbers, which run as follows:

1946: 30645001-30799737; Los Angeles 45000001-45002145

1947: 30799738-31011765; Los Angeles 45002146-45022452

1948: 31011766-31201086; Los Angeles 45022453-45041545

1949: 31201087-31242628; Los Angeles 45041546-45045426

Appraisal

There was no change in drivetrain for 1946–49, so everything we've said about driveability regarding the late prewar models applies to these as well. They weighed about the same, developed the same barely adequate 102hp, and had no suspension of transmission improvements except the standardization of front-wheel drive in 1947. The body styles to look for, in the order of desirability, are the convertible coupe, the Custom club coupe, the rather scarce seven-passenger Custom sedan, and the Town Sedan (preferable to the standard four-door sedan owing to scarcity and more formal appearance). Among those building on Dodge chassis was Superior Coach of Lima, Ohio, who offered many handsome funeral vehicles that are desirable among the "Professional Car" collectors.

Summary and Prospects

For the first time so far, we find certain Dodge models performing as investments better than a certificate of deposit over the past five or six years, the Custom club coupe returning 8 percent on your money, assuming it's a ninety-eight point show winner. As with the late prewar models, none of these Dodges showed any sign of taking a Ferrari-like plunge in value after the market doldrums of the early 1990s; their appreciation curve leveled out a little, but it didn't turn negative. Conclusion: these are genuine collector cars, not auctioncircuit widgets and grain futures substitutes; look for them to continue to gain in value by modest and reasonable amounts in the future: good buys, all.

Price History

95+ point condition 1	1982	1988	1995	return
Custom convertible	$9,500	19,000	22,000	7.2%
Custom club coupe	$3,300	6,500	8,500	8.2%
Custom 7-pass. sedan	$4,600	6,000	8,300	5.0%
Custom town sedan	$3,700	5,800	7,600	6.2%
Deluxe business coupe	$4,200	6,000	8,000	5.5%

Model Year Production

	1946	1947	1948	1949
Detroit	154,737	212,027	189,321	41,542
Los Angeles	2,146	20,307	19,093	3,881
Total	156,883	232,334	208,414	45,423
Percentage of 1946–49 Total	24.3%	36.2%	32.4%	7.1%

Estimating Production by Body Style: Dodge records only total 1946–49 production by body style. But by taking the above known percentages of model year production (based on serial number counts), we can make the following near-estimates of body style production by model year.

Body style	Total	1946	1947	1948	1949
DeLuxe 4dr sedan	61,987	15,063	22,439	20,084	4,401
DeLuxe 2dr sedan	81,399	19,780	29,466	26,373	5,779
DeLuxe bus. coupe	27,600	6,706	9,991	8,942	1,961
Custom 4dr sedan	333,911	81,140	120,876	108,187	23,708
Custom 4dr town sedan	27,800	6,755	10,064	9,007	1,974
Custom club coupe	103,800	25,223	37,576	33,631	7,370
Custom convertible	9,500	2,309	3,439	3,078	674
Custom 7 pass. sedan	*3,698	0	1,790	1,646	242
Custom limo prototype	2	2	0	0	0
Custom chassis	302	73	109	98	22

*Note: The seven-passenger sedan did not commence production until 1947, so for the purpose of an estimate I have apportioned the 1946 percentage equally between 1947 and 1948.

Specifications

Engine

Type: cast iron L-head six
Bore & stroke (in): 3.25 x 4.63
Displacement (ci): 230.2
Brake horsepower: 102

Chassis and Drivetrain

Transmission: three-speed manual, column-shift (1946) Fluid Drive standard (1947–48)
Suspension: i.f.s. with coil springs; beam rear axles with leaf springs

Measurements

Wheelbase (in): 119.5, seven-passenger sedan 137.5
Curb Weight (lbs): 3,150–3,350, seven-passenger sedan 3,800.
Tire size: 6.00x15 in 1946; 7.00x15 in 1947–49; seven-passenger sedan used 7.10x15 in 1946–49 inclusive.

Performance

Acceleration, 0–60mph: 22
Top speed (mph, approx.): 80–85
Fuel mileage (range): 13–19

Chapter 6

Fun	Investment	Anguish
7	6	3

Wayfarer Roadster

1949–52

History

Of all Dodge models in the early postwar years, the Wayfarer roadster has generated the most interest, I suppose because in its early 1949 form at least it was a true roadster, with side windows you had to take off and store, rather than roll down. Such quaint features were not very fashionable in 1949, of course, and Dodge switched to conventional roll-up windows on midyear Wayfarers. This made them in effect standard convertible coupes or cabriolets, but did not do anything to improve sales.

The Wayfarer's 115in wheelbase gave it the roadster's right proportions, and at $1,727 it was the lowest-priced softtop in America unless you count Crosley (most people didn't). It undercut Chevy's Bel Air convertible by $130 and Plymouth's ragtop by $200, and cost nearly $500 less than a softtop Dodge Coronet. At that price you'd think Dodge would have sold more than 5,400, but somehow the Wayfarer didn't click with the public. Sales tapered off to even more embarrassing levels in 1950–51, although in 1951 a price hike of some $200 must have put off a lot of potential buyers. The open Wayfarer was discontinued for 1952.

Dodge's underwhelming styling during this period probably didn't help, but the same boxy lines were used by Plymouth, whose ragtop sold 15,000 copies. Perhaps Dodge customers weren't convertible-conscious, since about 60,000 of them bought the more conventional Wayfarer coupe and two-door sedan in 1949. Undoubtedly the Wayfarer's image was wrong: the country was becoming performance-conscious, demanding longer, lower, wider and faster cars, and by the time the last Wayfarer roadster rolled off the Hamtramck lines, Dodge was making strong progress on its first V-8.

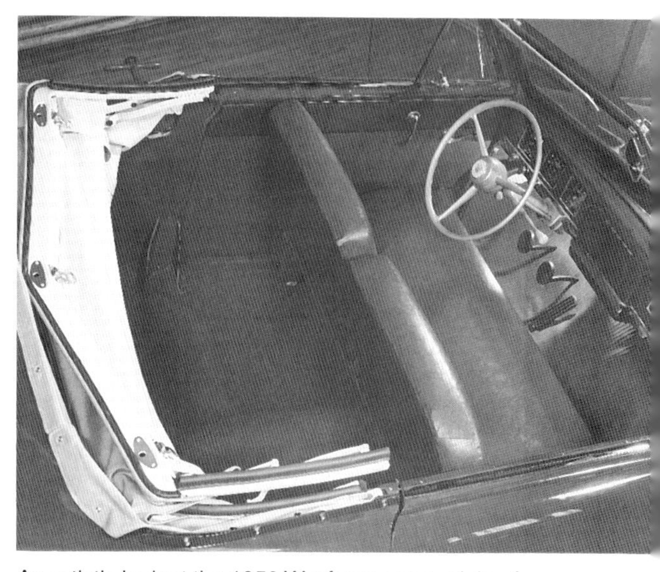

An artistic look at the 1950 Wayfarer, a very plain affair with no back seat, this one apparently equipped with no options at all: no radio, clock or even heater! Note the side braces behind the door, providing extra stiffening for the open body.

Artist's drawing of the 1949 Wayfarer roadster, the only one of the three-year run that was a true roadster (with take-out side windows).

The Wayfarer from a different angle emphasizes its nice proportions thanks to an extended deck. This appears to be an early '49 with side curtains. Whitewall tires would do it some good. Upright, fender-mounted taillights of senior Dodges were not used on this model.

For 1950 the Wayfarer received a cleaner grille, which did a lot to improve the looks of the front end, but was otherwise generally unchanged. Aluminum roof frame was not power-operated, but designed to go up and down with one hand.

Identification

All have "Wayfarer" script on front fenders. Removable side windows in chrome frames on early 1949 models only. Cross-hatched or egg-crate grille on 1949 models; double horizontal bar grille with central "bridge" containing Dodge badge on 1950 models; single-bar grille encompassing parking lights with individual Dodge letters and badge mounted on hood and shorter ram mascot with more curving horns on 1951 models.

Appraisal

It's not fast and it's no MG on the slalom course, and the interior is almost as spartan as a Willys Jeep, but the little Wayfarer is a cute rig that chugs along reasonably well given its short, light body and is reasonably fun to drive on a mild summer day. Don't try to track a well-driven Alfa into a decreasing radius turn, or run drags with '49 Fords and you won't be disappointed. I would strongly recommend looking for the genuine roadster (early 1949 only, with removable side windows), as this was the most interesting quality of the car, although it wasn't popular with the '49 public.

Expect 20mpg and don't push it over 60.

Summary and Prospects

The price guides, and the few ads I've seen for Wayfarers do not differentiate very much between the early '49, which I'm sure is the preferred model, and the later cars. Possibly this is because the '49 true roadster had much higher production than the later '49 and 1950–51 convertibles. If the price estimates are right, however, all four types of Wayfarer are good investments and I see no reason for their appreciation curve to change much in the years ahead.

That doesn't look like a very comfortable seat! Detail views of the 1950 Wayfarer emphasize the continuation of its spartan interior. At least this one has a heater installed . . . some dash areas that were woodgrained on higher price models remained painted on Wayfarers.

Larger taillights distinguished the rear of the 1950 Wayfarer, which was otherwise unchanged at the rear.

Wayfarer convertible made its last appearance in 1951, when it looked a lot more up-market thanks to a more streamlined hood and grille and, on this model, whitewall tires. The '51 had the lowest production, barely over 1,000, which makes it relatively more desirable, though the best investment is probably the early '49 genuine roadster.

Price History

95+ point condition 1	1982	1988	1995	Return
1949 roadster	$7,500	13,000	20,000	8.5%
1949 convertible	7,500	12,000	20,000	8.5%
1950 convertible	7,500	12,000	20,000	8.5%
1951 convertible	7,500	12,000	20,000	8.5%

Production

1949	1950	1951
5,420	2,903	1,002

Specifications

Engine

Type: cast iron L-head six
Bore & stroke (in): 3.25 x 4.63
Displacement (ci): 230.2
Brake horsepower: 103

Chassis and Drivetrain

Transmission: Fluid Drive; Gyro-Matic optional (1951)
Suspension: i.f.s. with coil springs, beam rear axle with leaf springs

Measurements

Wheelbase (in): 115
Curb Weight (lbs): 3,065 in 1949; 3,095 in 1950; 3,175 in 1951
Tires: 6.70 x 15

Performance

Acceleration, 0–60mph: 20
Top speed (mph, approx.): 80–85
Fuel mileage (range): 16–22

Chapter 7

Fun	Investment	Anguish
4	4	3

Sixes
1949–54

History

Dodge restyled for 1949 but took until April of that year to announce its new bodies. The look was Contemporary Chrysler: boxy, high, with not a lot of glass. Dodge advertised these cars as "Higher inside—lower outside, wider inside—narrower outside, larger inside—shorter outside," which confused everybody. More briefly, they looked unsleek. On the other hand, they had definite virtues: solid construction, quality materials, a powerful resistance to rust, reliable if not inspiring L-head cast iron six-cylinder engines. A gaudy crossbar grille on the initial run was replaced with cleaner horizontal dentures in 1950, when the Diplomat was added: Dodge's first hardtop, a year behind those of General Motors and good for 3,600 sales right off the bat. Gyromatic Drive, an improvement on standard Fluid Drive which eliminated gear changing, also appeared. Two basic model lines were offered, the D33 Wayfarer on a 115in wheelbase, and the D34 Coronet (including the stripped Meadowbrook sedan) on an expansive 123.5in frame. The extended wheelbase models had 14in more.

This basic arrangement lasted until 1953, when a restyle brought in a new line of cars on a shorter wheelbase, eliminating the stubby Wayfarer. Dodge introduced its first V-8 this year, and it immediately outsold the six-cylinder line. Horsepower was raised slightly in 1954.

Identification

1949: Egg-crate grille. 1950: Double bar grille with central "bridge" carrying Dodge medallion. 1951–52: Single bar grille encompassing parking lights with Dodge script and emblem mounted above grille on hood; shorter ram mascot with more curving horns. 1953: One-piece hood without center strip, one-

This is a 1954 Meadowbrook Six. Sixes were easily distinguished by their front ends, which bore a Dodge badge instead of a "V" badge. The '54 mascot had lost its bighorn sheep look and gone back to a design similar to that used in 1946-50.

Upright and boxy, the Coronet four-door sedan was top-of-the-line in 1949. Counting the Meadowbrook variation, which had a cheaper pinstriped interior instead of broadcloth, this body style accounted for about 145,000 sales.

The 1949 Coronet club coupe was better proportioned than the sedan, accommodated six, and sold for under $2,000; about 50,000 were produced.

piece curved windshield, blank upper grille cavity, lower grille has three horizontal bars, the top bar curving down at ends, the second and third forming a long oblong with six rectangular openings. All sixes carry a plain Dodge emblem on hood (different from V-8s which used a "V" emblem). Meadowbrook Special lacks chrome trim on windshield and backlight, has rubber floor mat. Meadowbrook carries trim on windshield/backlight and carpeted floor. 1954: single bar grille without vertical blades; bullet-top front bumper guards.

To differentiate the 1951 and 1952 models refer to serial numbers:

	1951	1952
Wayfarer, Detroit	37135001-37174917	37175001-37207644
Wayfarer, San Leandro	48008001-48009813	48009901-48011259
Wayfarer, Los Angeles	48506001-48507518	48507601-48508754
Coronet, Detroit	31663001-31867688	31867801-32038822
Coronet, San Leandro	45079001-45090487	45090601-45100113
Coronet, Los Angeles	45518001-45527382	45527501-45534770

Appraisal

Not entertaining behind the wheel, these Dodges, except for the Wayfarers, have more bulk to push around than the roadster discussed in the previous chapter, and not a lot of power with which to push it. The V-8 versions are far preferable, unless you're dealing with a convertible, which is desirable whatever engine it has. Note, however, that Dodge reserved the ragtop and hardtop strictly for V-8 models in 1953–54, leaving the six to push workaday sedans and wagons. Parts supply is reasonably good and mechanical reliability

Although Plymouth had introduced an all-steel wagon, Dodge's revival of this body style in 1949 was a woody, but only 800 were produced. Body colors were restricted to green, beige, and maroon. These are very handsome wagons when correctly restored.

The stubby Wayfarer coupe had a decent production run in 1949 but fell off fast in subsequent years. Note that Wayfarers do not carry the elaborate taillights of the Coronet and Meadowbrook. Dodge stylists didn't believe in hidden filler caps, either. . . .

The Coronet Diplomat for 1950 was Dodge's first entry in the popular "hardtop convertible" market and one of the more desirable early postwar collector models. The three-piece window was a nice touch. Only 3,600 were built, and this is a rare model today.

Dodge's last woody wagon was the 1950 Coronet, and just 600 were produced. A steel-bodied wagon was listed this year but at one hundred units was even scarcer.

high. Comparatively easy to restore, especially the 1949–52 models.

I prefer the '49–'50s to the later models: not only did they offer convertibles and hardtops, which were assigned to V-8s only in 1953–54, but they seem to me aesthetically more desirable. Their square styling has a nice vintage look to it, whereas the '51–52s seem more homely and the '53–54s kind of lumpy. On the inside, Coronets and even Meadowbrooks were handsomely trimmed in high-quality materials (Wayfarers were plainer), with a big mahogany grained dashboard containing nice, glitzy square instruments and chrome trim, which looks splendid if you fill in the blanks for radio and clock. The 1951 lost the woodgrain for a less successful leather-grain and received less flashy gauges and appointments. Models to look for aside from the obvious ragtop/hardtop: Coronet woody wagons (1949–50) and the rare Sierra wagons (1954), Wayfarer coupes and the neat 1949–52 two-door fastback sedan.

Summary and Prospects

Although not exciting cars, all these Dodges have performed well as investments over the past dozen years, with the Coronet convertibles and Diplomat hardtops in the lead. The latter doubled its value between the height of the market boom of the late 1980s and today, which is something you can't say of many Jaguars or Cadillacs, which were being touted as better than gold bullion by auction promoters not long ago. Again this points to the reputation and standing of plain-jane cars like these as "strictly for collectors." The hype merchants and investor crowd paid them no attention at all, and while the blue chip recommended buys of the eighties were leveling or falling in value, the boxy old six-cylinder Dodges just put on more. Safe, sane, and sensible investments they are, but that still doesn't make them any more fun to drive. One model which has not done particularly well since the late eighties is the early woody wagon, of which only 1,400 were built.

Although fastbacks were declining in popularity at the time, the 1951 Wayfarer was the most popular of that model, and accounted for over 70,000 sales. It's a much more interesting body style than the conventional coupe, and not hard to find or restore.

Dodge restyled around a new, more rounded and up-to-date body for 1953-54. The Meadowbrooks, like this '54 sedan, were detrimmed versions of the Coronets. Plain Dodge hood badge denotes a six-cylinder model—only a handful of Meadowbrooks had V-8s.

Price History

95+ point condition 1	1982	1988	1995	Return
1949–52 Wayfarer coupe	$3,800	5,500	7,500	5.8%
1949–52 Meadowbrook 4dr sedan	3,000	5,000	7,000	7.3%
1949–50 Coronet woody wagon	5,500	11,000	12,000	6.7%
1949–52 Coronet convertible	7,000	12,000	20,000	9.1%
1950–52 Coronet Diplomat hdtp	3,900	6,500	11,000	9.0%
1950–52 Coronet 8 pass. sedan	3,400	5,400	7,800	7.1%
1953–54 all models	3,200	5,500	7,000	6.7%

Production

	1949	1950	1951–2	1953	1954
Wayfarer coupe		9,342	7,500	6,702	0
Wayfarer fastback		49,058	65,000	70,700	0
Coronet/Mdwbrk. 4dr sedan	1	44,390	221,791	329,202	84,158
Coronet club coupe		45,435	38,502	56,103	*36,766
Coronet convertible		2,411	1,800	5,550	0
Coronet Diplomat hardtop		0	3,600	21,800	0
Cornet woody 4dr wagon		800	600	0	0
Coronet steel 4dr wagon	0	100	4,000	0	312
Mdwbrk./Coronet 2dr wagon	0	0	0	15,751	6,389
Coronet 8 pass. sedan		0	0	1,150	0
Meadowbrook club coupe					3,501
Meadowbrook 4dr sedan					7,894
Coronet club coupe					4,501
Coronet 4dr sedan					14,900

Specifications

Engine

Type: cast iron L-head six
Bore & stroke (in): 3.25 x 4.63
Displacement (ci): 230.2
Brake horsepower: 103; 110 in 1954

Chassis and Drivetrain

Transmission: Fluid Drive (1949–53), Gyro-Matic and overdrive
optional (1953–54), PowerFlite automatic optional (1954)
Suspension: i.f.s. with coil springs; solid rear axle with leaf springs.

Measurements

Wheelbase (in): 117
Curb Weight (lbs): 3,300–3,500
Tire size: 6.70x15

Performance

Acceleration, 0–60mph: 22
Top speed (mph, approx.): 80–85
Fuel mileage (range): 13–19

Chapter 8

Fun	Investment	Anguish
6	6	6

V-8s

1953–54

History

Its first V-8 in two decades, the 241ci Red Ram was a clear announcement that Dodge was in the performance game, and set an image that continues to this day. It was the ideal smallblock performance powerplant: compact in size, light in weight, capable of considerably more performance with the right tweaking. In concept it was a scaled-down version of the Chrysler hemi which had been introduced two years before; indicative of the performance emphasis on Dodge, it arrived two years before a V-8 Plymouth. Marketwise it was a canny move for Chrysler, a light, fast V-8 which sold for less than an Oldsmobile 88 or Mercury, Dodge's only Big Three rivals in the medium-priced field.

Emphasizing its fine new Red Ram V-8, Dodge fitted all '53 models with a prominent new "bighorn sheep" mascot and eights with a special scoop-type badge. Grille styling, new for 1953 (shown here), changed again for 1954.

Chrysler engineers had long been studying hemispherical head combustion chambers, and were now cashing in on what they had learned. The Hemi's advantages included smoother manifolding and porting, larger valves set farther apart, improved thermal efficiency, plenty of room for coolant passages, a more central spark plug location, and low heat rejection into coolant. Its main disadvantage was relatively high cost of manufacture.

Dodge wasted no time in proving both the economy and performance of the Red Ram V-8s. In the 1953 Mobilgas Economy Run a Ram scored 23.4mpg, while other V-8s broke 196 AAA stock car records at Bonneville. At El Mirage dry lake in California, Danny Eames drove another to a record 102.62mph. While Lincoln is famous for its victories in the Mexican Road Races of the fifties, few have noticed that Red Ram Dodges overwhelmed the Medium Stock class, finishing 1-2-3-4-6-9 in 1954.

Coronet's two-door sedan looked dowdy but offered Dodge's first performance to Ford in many years. Cost of the V-8 engine was only $114 more than the six, which was quite a bargain.

Looking much better with the optional continental type exterior spare tire and Kelsey-Hayes wires, the 1953 Coronet convertible started at $2,519. Loaded (this one even has the combination spotlight-rearview mirror and chrome wheel well moldings), it cleared $3,000. Only 4,100 '53 convertibles were built.

Behind the engine, the entire '53 drivetrain and chassis had been overhauled. Optional was Dodge's first automatic transmission, Gyro-Torque; underneath was the "Road Action" chassis, using double channel steel and a new suspension with big, flexible coil front springs. The body was new, too. Among the first cars styled by Virgil Exner, who had come over from Studebaker in 1949, the '53 with its one-piece windshield and lithe lines looked like a big step forward from the "three-box styling" of 1949–52.

For the '53 model year there were five V-8 models, all Coronets. The convertible, Diplomat hardtop and Sierra two-door wagon were more interesting, not least because of their 5in-shorter wheelbase than the four-door sedan and club coupe. This translated into weight savings, allowing the convertible, for example, to weigh little more than the four-door, despite its extra chassis and body stiffening.

The V-8s sold so well that Dodge brought out an upper-end Royal model in mid-1954, and although they arrived late, the Royals outsold the Coronets in the model year. Royals came as sedans, coupes, hardtops, and convertibles, while Coronets offered the same models plus three Suburban wagons.

The most desirable Dodge from this period is the luxurious limited edition, the Royal 500 convertible, named for the Indianapolis 500 race, which Dodge had paced that year. Selling for $2,632 including Kelsey-Hayes chrome wire wheels, an outside spare tire, special trim and a 150hp version of the Red Ram, it was the first Pace Car replica. Dealers could add a four-barrel Offenhauser manifold that must have made it a screamer, though Chrysler never quoted its actual horsepower.

Identification

1953: One-piece hood without center strip, one-piece curved windshield, blank upper grille cavity, lower grille has three horizontal bars, the top bar curving down at ends, the second and third forming a long oblong with six rectangular openings. Air vent with "V" emblem on hood. 1954: single bar grille without vertical blades; bullet-top front bumper guards.

Royal and Coronet carry identifying script on rear fenders; Royal has chrome fin on top of rear fender.

A new top-line Royal series was introduced for 1954, but sales were down and only 1,299 convertibles were built (along with fifty Coronet versions, the real rarity).

This 1954 Coronet factory photograph contains an element I can't explain: partially painted (instead of full chrome) headlamp rims. I have noticed this feature elsewhere only on the Indy 500 Royal pace car. Reader comment would be appreciated.

Appraisal

These little pocket-rockets won't knock your eyes out with razzmatazz styling, but they deliver remarkable performance for the money and are fun to drive by virtue of their compact dimensions, especially the more sporty models on the 114in wheelbase. Parts are not too difficult to find and service is reasonably straightforward. Production was pretty healthy in 1953, even for the usually-rare convertibles. It's much harder to find the rarer '54 models, the most desirable being the 500 convertible (701 built). Even scarcer, though, are the Coronet convertible (50) and hardtop (100).

Stylewise the '54 has a slight edge. Remember too that there was a Meadowbrook V-8 for 1954 only. Although this was the cheapest trimmed V-8 model, production was quite low: just 750 club coupes. Ordinarily I shy away form the obvious collector models like the 500, and recommend cheaper alternatives that may have been overlooked; in this case, however, I think the Royal 500 is a fine buy for the money, and I'd pay what extra it takes to get one in preference to a straight Royal or Coronet. (See below.)

Summary and Prospects

As we start moving into the performance model Dodges, the effects of the late-1980s "great collector car bubble" become more significant. Closed body styles and the relatively less exciting '53 Coronet convertibles have enjoyed nice steady gains, although not much in recent years; the Pace Car replica and its lookalike brother, the standard Royal convertible, have not gained at all in value since 1988, although a dozen years ago their prices were running well ahead of most medium-price field '54 convertibles. During the auction boom the 500 convertible was loudly trumpeted, and I've heard rumors of some being replicated out of standard 500s. But the bubble broke around 1990. At $13,000 for a very nice example, this is a very good price for a mid-fifties V-8 convertible, and with the bonus of high performance and limited production, the 500 is an excellent investment.

Price History

95+ point condition 1	1982	1988	1995	Return
1953–54 sedans, coupes, wagons	$3,600	5,500	6,500	5.0%
1953–54 station wagons	3,000	5,000	6,000	5.9%
1953 Coronet convertible	6,000	12,000	12,000	5.9%
1954 Royal hardtop	4,000	7,500	8,000	5.9%
1954 Royal convertible	7,500	12,000	12,000	4.0%
1954 Royal 500 convertible	9,000	13,000	13,000	3.1%

Production

	1953	1954
Meadowbrook coupe	0	750
Meadowbrook 4dr sedan	0	3,299
Coronet coupe	32,439	7,998
Coronet 4dr sedan	124,059	36,063
Coronet hardtop	17,334	100
Coronet convertible	4,100	50
Coronet Sierra/Suburban 2dr wagon	5,400	3,100
Coronet Sierra 4dr wagon	0	988
Royal coupe	0	8,900
Royal 4dr sedan	0	5,050
Royal hardtop	0	3,852
Royal convertible	0	1,299
Royal 500 convertible	0	701

Specifications

Engine

Type: cast iron hemi-head V-8
Bore & stroke (in): 3.44 x 3.25
Displacement (ci): 241.3
Brake horsepower: 140; (1953 and 1954 Meadowbrook); 150 (other '54s)

Chassis and Drivetrain

Transmission: three -speed manual with column-shift; overdrive and Fluid Drive optional in 1953; overdrive, Gyro-Matic and PowerFlite automatic optional in 1954.
Suspension: i.f.s. with coil springs; solid rear axle with leaf springs.

Measurements

Wheelbase (in): Four-door sedans and wagons and club coupes 119; hardtops, convertibles and two-door wagons 114.
Curb Weight (lbs): 3,400–3,700
Tire size: 7.10x15

Performance

Acceleration, 0–60mph: 16, less with Offy 4bbl manifold
Top speed (mph, approx.): 95–100
Fuel mileage (range): 13–18

Chapter 9

Fun	Investment	Anguish
8	8	4

Flair Fashion

1955–56

History

A young designer named Maury Baldwin, working under Virg Exner, designed the 1955 Dodge and Plymouth, which are thus rather different in form from the '55 DeSoto, Chrysler and Imperial, which received more of Exner's personal attention. I interviewed Baldwin twenty years ago; though it was then twenty years since the design project ended, he still had fresh memories of the experience, which was an exciting one: Virgil Exner had set out to change Chrysler's image from arch-conservative to style leader, setting the pace even for General Motors. And by 1957, he would succeed.

The '55 Dodge was as fresh as anything around in that great sales year. Although Dodge remained eighth in sales, where it had been since 1953, production doubled from the year before: only the third year to date when Dodge had built more than 300,000 cars (the others were 1950 and 1951). Baldwin's clean, balanced lines, split grille, imaginative two- and three-tone color schemes and asymmetrical, aircraft-inspired cockpit combined with a longer wheelbase and sharp interiors to form the prettiest Dodge in a generation. The new, 270ci Hemi V-8, with up to 193hp, took care of performance.

As from its introduction, the V-8 dominated sales. While sixes were offered only on Coronet sedans and wagons there was a broad line of Coronet, Royal, and Custom Royal V-8s, including three hardtops. In 1956, tailfins were added and the Powerflite controls went from a dashboard lever to pushbutton console mounted left of the steering wheel. Body styles were the same, with the addition of four-door hardtops in all three V-8 lines. Top gun V-8 was the D500, which developed 260hp at 4400rpm from the new 315ci engine. This potent engine could be ordered in any model, including an otherwise ordinary Coronet club sedan: a veritable Q-ship at stoplight grands prix. Official results were also good: in 1956, Dodges won eleven Grand National stock car races, and a Custom Royal sedan set over 306 speed records at Bonneville.

One mid-1955 trim package which would later be derided by our cultural chauvinists was La Femme, a Custom Royal Lancer usually painted Heather rose and Sapphire white, gratuitously equipped for the "woman driver." It came with a color-keyed cape, boots, umbrella and shoulder bag, and floral upholstery fabric. It was, of course, a male-stylist's idea of what women wanted in a car, and there was even a he-man counterpart, the Coronet Texan, with a more macho trim package. It isn't fashionable to say so, but let's get real: in 1955 La Femme was probably just what a lot of car-conscious women wanted. It was a big hit at the auto shows and undoubtedly added to dealer floor traffic.

For midyear 1956 La Femme returned in a lavender and white color scheme, with a

The most striking Dodge in decades, Maury Baldwin's '55 introduced three-tone color schemes (along with Packard and DeSoto) and unprecedented glass area. It was hard to compare this model with the Dodge that had broke all earlier records just five years before. The "Flair Fashion" '55s ushered in a new era of high style and performance, a tradition that continues forty years later. Top-line hardtop was this Custom Royal Lancer; a show-winning example cost about $13,000 in 1995.

One of Baldwin's more novel ideas was the split-level, two-toned hood, which gave Dodge a totally different appearance from its Plymouth cousin, which shared the basic body shell; this was the beginning of the end for Dodge hood ornaments.

similar interior featuring gold flecked fabrics; its accessories were now a color keyed umbrella, rain hat and purse hook. The Texan was back too, along with a new and ultra-exotic Dodge, the Golden Lancer. This fast and fancy flyer was a D500 hardtop finished in gold and white, matching the style of the '56 Plymouth Fury. Inside, it had gold finished instruments, windshield and door moldings, and special upholstery mixing whites, blacks, and greys. I've often thought that one of these, mated with a Fury, DeSoto Adventurer, and Chrysler 300B, all in white trimmed with gold or saddle tan, would make a fabulous collection, redolent with the style and performance of the middle-fifties.

Identification

As a rule, the 1956 Dodge rear fenders had prominent, built-in tailfins, while the '55s didn't. The exceptions were the '55 Custom Royals, which carried small, bolt-on, chrome plated fins; and the '56 station wagons, whose

Royals came only with V-8 engines, and four-door sedans were by far the most popular; $7,500 buys a top-condition example today.

fenders were carried over from 1955 but fitted with bolt-on chrome plated fins. (Its low volume didn't justify tail-finning the 1956 wagons to match the other bodies.) Both model years used a divided grille and had a raised center hood section leading with a chrome-capped, simulated airscoop carrying one vertical blade in 1955, two in '56. Note: in 1955 the term "Lancer" was used on two Custom Royals that were not hardtops: the convertible coupe and an up-market four-door sedan. In 1956 "Lancer" referred to two- and four-door hardtops only.

Appraisal

Three-tone paint jobs look gaudy nowadays and tailfins are definitely a nostalgia item. Still these Dodges look good, especially

Convertibles in showworthy condition are $20,000-plus cars today, closer to $30,000 with the D-500 engine. A nice touch on this '56 Custom Royal is the fabric Dodge badge sewn into the top quarter.

Dodge's first four-door hardtops were offered in 1956. These airy sedans have not matched two-door hardtops in price appreciation, and may be bought for little more than standard sedans.

The 1956 Royal Lancer was a typical up-market model. Two-door hardtops have had excellent appreciation in value over the past five years and are the best buys next to convertibles.

Dodge wagons came with two and four doors and with two and three bench seats. They were distinguished by "tacked-on" chrome tailfins rather than the built-up steel fins in 1956. Their prices today run even with those of sedans.

the '55. I've often scanned the pages of *Hemmings Motor News* ogling '55 Custom Royal convertibles, routinely finding nice ones but always deciding I couldn't afford one. If I had my choice, that would be my Dodge. Maury Baldwin was a friend of mine.

The most desirable Dodges are the D500s (any model and body style, but especially hardtops and convertibles), the La Femme and Texan, and of course the fabulous Golden Lancer, a rare car destined to be owned by a fortunate few. Bargain collectibles include the Coronet and Royal hardtops and the Lancer four-door hardtop (available in three models in 1956). The cheapest ragtop is the Coronet convertible, offered in 1956 only. Sixes, though much scarcer than V-8s, are less desirable (all convertibles are V-8s). Mechanical parts are in fair supply, body parts hard to come by. The cars are mechanically straightforward, easy to work on, solid and reliable.

The most interesting accessory to look for is Highway Hi-Fi (introduced 1956 but most likely to be found on the '57s), which played special records with a needle designed to stay in the grooves regardless of road jounce. After a few years of low sales it was withdrawn, its failure put down to cost, limited choice of music and the short duration of play. Today's CD players are the modern derivation of Chrysler's idea from nearly forty years ago.

Summary and Prospects

Even if you bought a '55 or '56 Dodge at the height of the investment bubble in 1988, you still made out. At that time some models had doubled their value of six years before. What's remarkable is that they've done at least as well in the *last* six years, fine examples doing far better than any certificate of deposit. Even a plug ordinary two-door Coronet has appreciated decently. They were utterly unaffected by the recession of the early 1990s. This seems to have occurred because (a) mid-1950s Dodges didn't excite the auction/investment types to the extent that later Dodges like Daytona Hemis, and (b) they were still undervalued in 1988. Now that you and I have discovered them, what happens? People will probably start paying more attention. I believe, though, that they'll continue to appreciate handsomely.

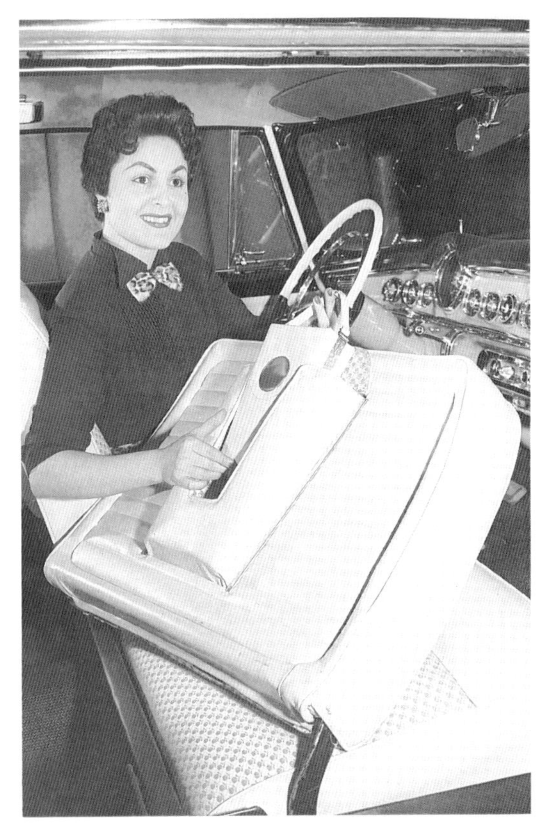

"La Femme" was a special Custom Royal targeted at female buyers, painted lavender and white for 1956 and fitted with removable handbag and umbrella. A best buy among hardtops, it is one of the few fifties trim options in reasonable supply on today's collector market; still, you don't often see them offered for sale.

Price History

95+ point condition 1	1982	1988	1995	Return
Sedans and wagons	$3,500	5,500	7,000	5.9%
1955 Royal 2dr hardtop	3,700	7,500	12,000	10.3%
1955 Custom Royal convertible	7,400	11,000	23,000	9.9%
1956 Coronet convertible	7,000	9,500	23,000	10.4%
1955–56 La Femme hardtop	4,000	9,000	18,000	13.3%
(Add one-third for any 1956 model with D500 engine.)				
1956 Golden Lancer D500 hardtop	5,000	10,000	20,000	12.2%

Production

1955 Coronet six: 15,976 4dr sedans; 13,277 2dr sedans; 3,248 Suburban 2dr wagons; 1,311 Suburban 4dr wagons for six or eight passengers.
1955 Coronet V-8: 30,098 4dr sedans; 26,727 Lancer 2dr hardtops, 10,827 two-door sedans; 4,867 Suburban 2dr wagons; 4,641 Suburban 4dr wagons for six or eight passengers.
1955 Royal V-8: 45,323 4dr sedans, 25,831 Lancer 2dr hardtops; 5,506 Sierra 4dr wagons.
1955 Custom Royal: 55,503 4dr sedans; 30,499 2dr hardtops; 3,302 Lancer convertibles.
1956: Dodge does not break out production except by model totals: 142,613 Coronets; 48,780 Royals; 49,293 Custom Royals.

Specifications

Engine

Type: cast iron L-head six and ohv V-8
Six: 230.2ci (3.25 x 4.63), 123hp, 1955; 131hp, 1956
V-8s: 270.1ci (3.63 x 3.25), 175hp, 1955 Coronet & Royal; 183–193hp,
1955 Custom Royal; 189hp, 1956 Coronet
315ci (3.63 x 3.80), 218hp, 1956 Royal & Custom Royal; 260hp, optional on all 1956s

Chassis and Drivetrain

Transmission: three-speed manual, column-shift; over drive or PowerFlite automatic optional. Pushbutton Power Flite in 1956
Suspension: i.f.s. with coil springs; solid rear axle with leaf springs

Measurements

Wheelbase (in): 120
Curb Weight (lbs): 3,300–3,700
Tire size: 6.70x15 (six), 7.60x15 (1956 Custom Royal), 7.10x15 (other V-8s)

Performance (V-8s)

Acceleration, 0–60mph: 11–15
Top speed (mph, approx.): 95–105
Fuel mileage (range): 12–18

Chapter 10

Fun	Investment	Anguish
8	8	7

Forward Look
1957–59

History

Chrysler's Forward Look sent the rest of the industry scrambling to catch up, notably General Motors, caught unprepared in the third year of its three-year styling cycle. For the first time since styling became an automotive art (repeated, in my opinion, right now in the 1990s), Chrysler Corporation design led the league. It had come from nowhere: five years before, Chrysler cars had been the dullest looking things on American wheels.

As dramatic as the 1955-56 models had been, the 1957 was a revelation. Now Dodge and its corporate brethren commanded styling leadership for the entire industry, on the swept-wing verve of Virgil Exner's "Forward Look." Fenders had risen to hood height the '57 had unprecedented power. It was the most impressive Dodge of the decade.

Longer, lower, wider, aggressively tailfinned, trailed by huge V-8s, and belting out 345 gross horsepower, the Swept Wing 1957–59 Dodge won hearts but broke no production records. The '57 was the most successful, with calendar year production approaching 300,000. But 1958 was disastrous. Production plunged by more than 50 percent, barely enough to outsell Cadillac. Dodge rebounded with a restyled '59 created out of the same basic materials and built 156,000 for its usual eighth place finish that year, but this was still shaky compared to its success in the banner years of 1955 and 1957.

Today these cars represent Dodge's contribution to the so-called age of excess, packing outrageous styling and accessory gimmicks with blinding acceleration and awful gas mileage. But they also made certain solid contributions to automotive technology: torsion bar front suspension, dramatic lowness, unprecedented glass area.

In 1957 as before, Dodge chose not to follow its brother divisions with a limited edition supercar, but to offer D500 equipment on anything from a base Coronet to a Custom Royal convertible. All D500s were equipped with extra-stiff shocks, springs and torsion bars, providing what *Motor Trend* called "close liaison with the road." Firm suspension put them at the top of their class in handling, and with the 245hp engine, the '57 ran 0–60 in under ten seconds. The D500 was continued in 1958–59,

The Custom Royal Lancer carried a 325ci Red Ram V-8 as standard and offered performance up to the 310hp D-500 engine. Custom Royals can instantly be identified by their distinctive grille teeth (see preceding photo).

A rare monotone paint job on a Custom Royal Lancer convertible photographed outside the factory. It's hard to believe that one like this is now worth ten times its 1957 price. With D-500 equipment, this is my nominee as the most desirable Dodge of the fifties.

Quad headlamps and a new grille distinguished the 1958 model. Lack of grille teeth identify this convertible as a Coronet model. Scarcer than the Royal Lancer ragtop but no less costly on the collector market, it's a hard Dodge to find.

its expensive Hemi engine replaced with a wedge-head V-8. Offered with optional fuel injection for 1958, the 361 wedge produced 333hp at 4800, the highest of any Dodge in history.

Dodge continued to use its 122in 1957 wheelbase in 1958–59, giving the cars mild facelifts, standard quad headlamps and a less massive front end. The model line-up was generally unchanged, though a top-of-the-line Regal Lancer hardtop appeared in mid-1958. Several interesting new accessories premiered including Co-Pilot, an early cruise control, (1958) and the swivel front seat, which pivoted outward as the front doors opened (1959); Highway Hi-Fi continued as an option.

Identification

The 1957 is unmistakable, with its huge, single-bar grille dipped in the middle under a wide Dodge emblem, twin windsplits on the hood and headlamps and parking lamps incorporated under the headlamp brows. The 1958 grille bars are truncated, leaving space in between for a cross-hatched grille; '58s also carry three windsplits on each side of the front bumper. The 1959 model is very busy looking, with a fine mesh grille behind stubby grille bars and the headlamp "eyebrows" slanting downward toward the center.

Appraisal

The fatal flaw of these Dodges is their propensity to rust, especially in places that are hard to inspect: the undersides of those big headlight brows, deep inside the finned rear fenders, as well as in the more obvious places, like fenders and rocker panels. This is a major hazard and suggests you look for a car from a rust-free area, or be prepared for lots of bodywork and repairs which may only prove temporary. Fuel injection was a troublesome and therefore shortlived innovation. Few buyers opted for it so you probably won't encounter

Owing to the concurrent recession '58s are much harder to find than '57s, especially desirable models like this Custom Royal hardtop. Lately two-door hardtops have been gaining in value faster than any other body style of 1957-58 Dodge.

The limited edition Regal Lancer is a very scarce trim variation painted bronze and black, with a unique side molding treatment and custom interior including bucket-seats, deep foam rubber seating, and "Acousti-Foam" headliner. Figure such a car to be worth 25 percent more than comparable standard Lancers.

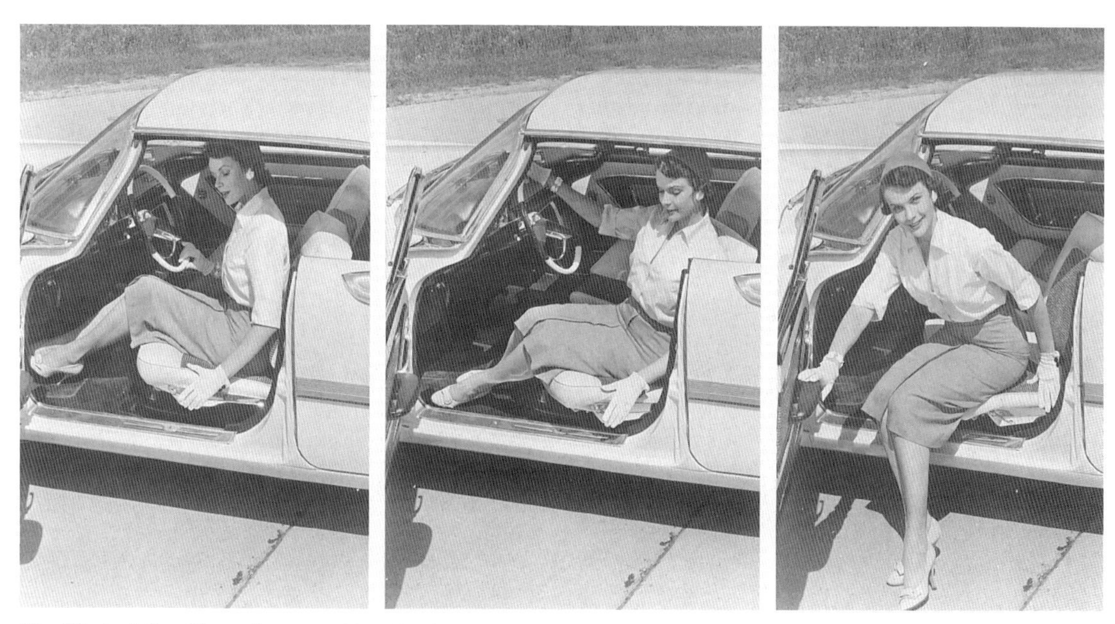

The "Swivel-Seat" in action on a 1959 Dodge. It's curious why this feature wasn't more popular; collectors certainly enjoy it nowadays.

On three-seat Dodge wagons, the rear seat was set observation-style facing the rear. This photograph was taken at the banked track at Chrysler's Chelsea, Michigan, proving grounds.

it, but it should be avoided in any case, since parts and service pose big problems.

Summary and Prospects

The sportier models of these Dodges defy all the trends of the collector car market over the past five or six years. Custom Royals, which had increased only modestly in value from 1982 to 1988, suddenly took off, convertibles joining the ranks of high priced "neoclassics," running to $30,000. Prime condition two-door hardtops did better, in terms of compound rates of return, than any pre-1960s Dodge. One caveat: these are price guide figures, not real world figures, and if I judge the asking prices properly, 1957–59 Dodges do not quite command what the price guides say they do at the time of writing. I think they are all solid investments for the future, however. In many ways the Swept-Wing Dodges are unique artifacts of an era, unimitated either by their predecessors or successors.

Last year for the three-year styling run of the "Forward Look" Dodge, the 1959 model shaved the hood emblems, kept the tailfins, and adopted an even busier grille. There's a hair less interest in these than in the 1957-58, a judgment that is almost certainly based on aesthetics.

Price History

95+ point condition 1	1982	1988	1995	Return
Sedans & wagons	$2,000	4,500	6,000	9.5%
Custom Royal convertible	6,200	9,000	30,000	14.0%
Coronet convertible	5,500	8,000	25,000	13.4%
Custom Royal 2dr hardtop	3,000	6,000	18,000	16.0%
Royal 4dr hardtop	2,500	5,000	7,500	9.5%

Production

	1957	1958	1959
Coronet	60,979	77,388	96,782
Royal	40,999	15,165	14,807
Custom Royal	55,149	23,949	21,206
Regal Lancer	0	1,163	0
Station Wagons	30,481	20,196	23,590

Specifications

Engines

Types: cast iron L-head six and ohv V-8
Six: 230ci (3.25 x 4.63), 138hp
V-8s: 325ci (3.69 x 3.80), 245–260hp, all 1957s; 252–265hp,
1958 Coronet & Royal; 285–310hp, optional 1957 D500
326ci (3.95 x 3.31), 255hp, 1959 Coronet
350ci (4.06 x 3.38), 285hp, 1958 Custom Royal
354ci (3.94 x 3.63), 340hp, optional all 1957s (D500)
361ci (4.12 x 3.38), 305–333hp, optional all 1958s (D500), 1959 wagons, Royal & Custom Royal
383ci (4.25 x 3.38), 320hp, optional all 1959s (D500); 345hp, optional all 1959s (Super D500)

Chassis and Drivetrain

Transmission: three-speed manual, column-shift; PowerFlite and TorqueFlite automatic transmission optional
Suspension: i.f.s. front torsion bars; solid rear axle with leaf springs

Measurements

Wheelbase (in): 122
Curb Weight (lbs): 3,200–4,000
Tire size: 7.50 x 14 (Coronet), 8.00 x 14 (other models & Coronet convertible)

Performance (D500)

Acceleration, 0–60mph: 8.5 to 9.5 seconds
Top speed (mph, approx.): 120mph

Chapter 11

Fun	Investment	Anguish
7	7	6

Big Darts
1960–62

An early example of safetyengineering? No, it's a Dodge "plastic person," used to measure interior dimensions in designing the 1960 models. Exner's asymmetrical front seat, with a higher backrest for the driver, is also visible in this photo.

History

Named after a Chrysler Corporation showcar, the first Dodge Dart was a well-timed piece of market strategy which, at first, dealt handily with changing customer preferences. By restyling, cutting 4in off the standard size wheelbase, and introducing the fine new 225ci Slant Six, Dodge created a kind of "intermediate," two years before that term was in common usage. A full line of six and V-8 Darts in three trim stages (Seneca, Pioneer, Phoenix) came in all the usual body styles, station wagons retaining the longer 122in wheelbase, and Dart prices started almost $200 below the lowest priced 1959 Coronet.

The Dart lifted Dodge back to its accustomed sixth place in sales and helped set a new model year record of 368,000. But the boost didn't last: in 1961, with Buick, Oldsmobile, Mercury, and Pontiac all offering compacts and economy versions of their larger cars, the Dart's competition multiplied, and the Lancer (Dodge's version of the Plymouth Valiant) was a sales disappointment too. The following year, with Ford's Fairlane and the Chevy II added to the field, was worse yet, though not as bad as 1958. It wasn't until Dodge put the Dart name on a handsome new compact which replaced Lancer in 1963 that the Division's fortunes picked up again.

The Dart (and its senior linemate the Polara) introduced Dodge's first unit body-chassis, a single, all-welded body that replaced the

Phoenix, the most luxurious 1960 Dart, carried the 318 V-8 as standard. Those four-flipper wheel covers were hot items at midnight auto sales, sought after by hot-rodders and customizers to make their wheels glitter in motion.

traditional bolted-together separate body and frame. Signature features like torsion bar front suspension ("Torsion-Aire Ride") and optional swivel seats continued. Dart V-8 engines spanned a wide gap from the standard 318 through the potent 383 with 325hp. The D500 this year carried this engine with dual four-barrel carburetors and ram induction, pumping out 330hp. This package was confined to the top-of-the-line Dart Phoenix.

A facelift gave the '61 Dart a full-width, concave grille incorporating the headlamps and odd, reverse-slant tailfins. The triple lineup of Seneca, Pioneer, and Phoenix sub-models continued, the Phoenix again available as a D500 with the formidable ram-induction 383 V-8. In mid-1961 a 413ci wedge V-8 was added for even more performance. These fire-breathing Darts followed the successful concept of the earlier Olds 88: a light body with a state-of-the-art V-8 giving terrific performance. At their best, they delivered almost 1hp for every ten pounds of weight. Big-engined Senecas were the darlings of hot-rodders and stock car racers, and ram-induction Darts were top eliminators on the dragstrips. Ram-Charger 413s routinely turned in under-thirteen-second quarter miles, and held four National Hot Rod Association records. In 1960, a supercharged 383 Dart with tall gearing hit over 190mph at Bonneville!

A Phoenix at work at the Chelsea Proving Grounds. One of the smallest Dodges to date with torsion bar front suspension, Dart was known for its roadability and handling precision.

For 1962 the Dart received a new body on a shorter wheelbase, inspired by Virgil Exner, who had shorn the tailfins that won him fame and nicknamed these cars the "plucked chickens." The Seneca-Pioneer-Phoenix submodels were replaced by base Dart, 330 and 440, and the Polara name was transferred from the senior Dodge to a top-of-the-line Dart called Polara 500, sold in hardtop or convertible form only. Fitted with bucket seats and center console in the fashion of the times, the Polara 500 was readily identifiable through its partly blacked-out grille; the convertible version was the smartest looking Dodge of its generation.

This time, though, Exner was too far ahead of his time, and a downsized Dart was not welcomed. While most manufacturers increased their sales this year, Dodge dropped. Even the wedge-head 413 V-8 didn't help. Dodge reacted fast, dumping the "plucked chicken" after just one year, increasing its standard wheelbase and slapping the Dart name on a new compact. The result was a brisk recovery.

Identification

The 1960 Dart had a grille composed of vertical bars and tailfins that ended a few inches ahead of the rear fenders, which tapered to round taillamps. The '61 was severely facelifted, with a cleaner, concave grille incorporating the headlamps, a simpler front bumper, and reverse slant tailfins which rose over the rear wheel arches and tapered down toward the tail. The '62 lost both the wrap-around windshield and the tailfins, sprouted Valiant-like side ridges on its fenders, and had a simple grille housing the inboard headlamps; the outer headlamps were at the extremities of the fenders.

Appraisal

While Chrysler's Unibody construction brought unprecedented rigidity and strength to automotive bodies, it also suffered from the traditional unit body-chassis malady: susceptibility to the dreaded tinworm. It is therefore important to avoid any Dart that shows signs either of extant rust or efforts to eliminate it. Ram induction engines present parts problems, are hard to tune; all 361, 383, and 413 V-8s are not really happy on the swamp water that passes for high-octane gas these days, so you may experience driveability problems and be unable to eliminate them. Models to look for: the early Phoenix convertibles, which are worth much more than any other Darts; the Polara 500, which had a relatively low production run; the Slant Six Phoenix, a rarity that still performs decently.

Summary and Prospects

Peculiar styling, a tendency to rust out from under you and impossibly thirsty big V-8s that are hard to tune: is there anything good about the original Dart? Lots of things. Start with performance: these cars were among the fastest U.S. production automobiles of their

Not everybody liked the '61, especially in detrimmed form, like this stripped Seneca four-door. The scooped-out grille and reverse slant rear fins just looked odd, and the latter belied everything Chrysler had been saying about fins adding stability at high speed. The following year they'd be gone for good.

Dodge offered twenty-nine models in the 1961 Dart range, one of the best of which was the Phoenix two-door hardtop. Starting at $2,700, the typical Phoenix cost about $3,500.

time. Or take value: convertibles and hardtops cost half the price of their mid-fifties counterparts. Light weight assures good performance even with the 318 or Slant Six engines, and the styling is oddly attractive in some cases, such as the '62 Polara 500. In 1982, where we begin our price survey, they were just starting to appreciate in value after two decades' depreciation. Today they look like reasonably good investments though I shouldn't expect the rate of increase that we've seen over the past decade.

Pioneer was the midrange model in 1961; wagons were available with the new Slant Six but are most often found with V-8s.

For 1962 the Dart was shorn of tailfins and dramatically reduced in size. Front end styling was radically different, the two pairs of headlamps set within and without the grille. It was an early attempt at reducing the bulk of standard American cars, but it was premature. The 440 four-door hardtop and nine-passenger wagon are shown.

Specifications

Engines

Types: aluminum block ohv slant six & cast iron ohv V-8
Six: 225ci (3.40 x 4.13), 145hp standard
V-8s: 318ci (3.91 x 3.31), 230hp standard; 260hp optional. 361ci (4.12 x 3.38), 295hp, optional 1960 Pioneer & Phoenix; 305hp (D500) optional all 1961–62s, standard 1962
Polara 500–383ci (4.25 x 3.38), 325hp, optional 1960 Phoenix; with ram in duction 330hp (D500), optional 1960 Phoenix & all 1961s 413ci (4.25 x 3.75), 375hp, optional mid-1961; 410–420hp, optional 1962

Chassis and Drivetrain

Unibody construction
Transmission: three-speed manual, TorqueFlite automatic optional, Powerflite automatic optional 1960 only.
Suspension: torsion bar front suspension, solid rear axle with leaf springs.

Measurements

Wheelbase (in): 1960–61: 118, wagons 122. 1962: 116
Curb Weight (lbs): 3,300–3,500
Tire size: 1960–61: Six, 7.00 x 14; V-8s, 7.50 x 14; Wagons, **8.00 x 14; 1962:** 8.00 x 14; Wagons 8.50 x 14

Performance (383/413 V-8s)

Acceleration, 0–60mph: 9–11
Top speed (mph, approx.): 110–125
Fuel mileage (range): 10-15

Price History

95+ point condition 1	1982	1988	1995	Return
1960–61 Phoenix convertible	$2,600	7,500	16,000	16.3%
1962 Polara 500 convertible	2,400	6,500	11,000	13.5%
1962 440 convertible	2,200	6,000	10,000	13.4%
1962 Polara 500 2dr hardtop	1,800	4,500	6,500	11.2%
1960–62 other hardtops	1,500	4,000	5,500	11.4%
Other models	1,000	3,500	5,000	14.3%
(Add 10–25% for ram-induction V-8s)				

Production

	1960	1961	1962
Seneca/base six	93,167	60,527	43,927
Pioneer/330 six	36,434	18,214	11,606
Phoenix/440 six	6,567	4,273	3,942
Seneca/base V-8	45,737	27,174	17,981
Pioneer/330 V-8	74,655	39,054	26,544
Phoenix/440 V-8	66,608	34,419	42,360
Polara 500 (116in wb)	0	0	12,268

Chapter 12

Fun	Investment	Anguish
4	4	6

Polara/Matador
1960–61

History

Dodge's only "full-size" models at the outset of the sixties, Polara and Matador captured a diminishing share of Division sales, just 40,000 for 1960 and less than half that number in 1961. (The 1962 Polara 500 was an up-market Dart, covered in the previous chapter.) Matador and Polara were readily identifiable from the Dart in 1960, with a longer rear end topped by more abrupt tailfins and a more elaborate grille. The following year the Matador was dropped (its name was picked up later by American Motors), leaving Polara the only Dodge (besides Dart wagons) on the long, 122in wheelbase. For '61, Polara adopted the concave Dart grillework with a big star-like emblem to differentiate it. Cylindrical taillights were cupped in the outer edges of the rear fenders (they were also used, by necessity, on Dart wagons). When the Polara was shoved over onto the Dart chassis in 1962, its place was taken by the Chrysler-based Custom 880 (see Chapter 14).

Matadors started at just under $3,000, well up from the Darts, and Polaras began at $3,200. As usually optioned, they cost upwards of $4,500. The usual body styles of sedan, two- and four-door hardtop and station wagon were offered, plus a Polara convertible. Whereas Polara wagons were four-door hardtops, however, the Matador was conventional with a solid "B" pillar, and sold for about $400 less than its senior counterpart.

Nothing smaller than the 361 V-8 was ever offered on these cars, and they could be optioned to the max, which usually meant 330hp from the 383, although a 413 V-8 was available on mid-'61 Polaras. Essentially they were the direct continuation of the standard-sized generation that began in 1957, but they didn't suit the tastes of the times as well as the Dart. Dodge Division rationalized the whole

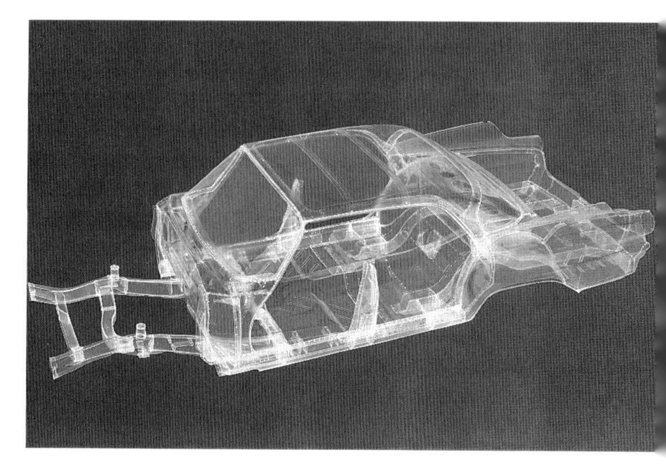

Nineteen-sixty brought a fresh set of Dodge model names and an entirely new construction method: "Unibody," with the body/chassis comprising a single unit, dramatically portrayed in this "ghosted" photo of a Polara/Matador sedan.

The big Polara, with its truncated tailfins and glitzy chrome trim, was Dodge's top-line product for 1961, selling for close to $4,000 as usually equipped. Partly as a result, it sold in small quantitites. Only a few thousand of these two-door hardtops were built, with most Polaras being four-door hardtops and sedans.

size problem in 1963 by going to a middling 119in wheelbase for all its cars except the compact Lancer and the Custom 880.

Identification

In 1960, tailfins were shorter than on Darts, leaving an extended rear fender carrying round taillights. Polaras had a chrome gravel shield on rear fenders behind wheel wells, while Matadors were painted in this area and carried a narrow chrome ornament on the front fender. The '61 Polara carries a four-pronged star emblem in the center of its Dart-based grille and cylindrical taillamps cupped in the rear fenders, where they are clearly visible from sides as well as rear. There was no Matador in 1961.

Appraisal

These are essentially heavy Darts, larger and slower, but more luxurious, Polaras especially. Because they sold primarily for luxury and not performance, they are less apt to be fit-

The 1961 Polara four-door. The following year, Polara was shifted to the Dart bodyshell, while Dodge's "full-size" product became the Custom 880.

High performance and smooth good looks were characteristic of the desirable 1962 Polara 500 convertible, powered by a 305hp V-8 and equipped with bucket seats and blacked-out grille.

ted with the really high performance engines. Problem areas affecting Darts, including susceptibility to corrosion and the routine difficulty of big engines designed to run on gas no longer available apply to Polaras and Matadors as well. An additional concern is a great scarcity of body and trim parts. The '61 Polara was a smoothly styled package that looks good on its long wheelbase, and the Polara four-door hardtop station wagons are worth looking for. But it's less fun to drive and not as rapid off the line as a contemporary Dart with the same engine.

Summary and Prospects

Despite what the writer thinks about them, Polaras and Matadors have done extremely well over recent years, and if you invested in a ninety-five-point example a decade ago you certainly can't complain about what it's worth now. Remember, though, that our Price History is for "condition 1" cars, the cream of the crop; those in poorer condition haven't done as well. Still, compared to other Dodges the Polara and Matador look like excellent investments now and in the future. One point worth noting: there's a big difference in the value collectors place on the two Polara convertibles. The '60 is clearly more desirable, despite its higher production, probably because with those soaring tailfins it's a better example of the "neoclassic" age than its more conservative and refined successor. The '61 convertible, on the other hand, is worth more than the downsized Polara 500 (see previous chapter), and has appreciated almost as much as the '60.

The biggest surprise is the lowly Matador hardtop, which has certainly been a blue chip stock for car investors, no doubt because it was severely undervalued ten or twelve years ago.

Price History

95+ point condition 1	1982	1988	1995	Return
1960 Polara convertible	$3,000	7,500	19,000	16.5%
1960–61 Polara 4dr hardtop wagon	1,500	3,600	6,000	12.2%
1960 Matador 2dr hardtop	1,600	5,000	13,000	19.6%
1961 Polara convertible	2,600	6,500	13,000	14.3%
Sedans & Matador wagons	1,000	4,000	5,500	15.2%

Specifications

Engines

Type: cast iron ohv V-8

V-8s: 361ci (4.12 x 3.38), 295hp, standard Matador; 265hp, standard 1961 Polara 383ci (4.25 x 3.38), 325hp, standard 1960 Polara, optional 1961 Matador & 1961 Polara; 330hp optional all models 413ci (4.25 x 3.75), 375hp, optional mid-1961

Chassis and Drivetrain

Unibody construction

Transmission: three-speed manual was ostensibly standard but two-speed Powerflite automatic (1960) or three-speed TorqueFlite automatic (1960–61) were almost always equipped

Suspension: torsion bar front suspension, solid rear axle with leaf springs

Measurements

Wheelbase (in): 122

Curb Weight (lbs): wagons 4,000–4,300; other models 3,700–3,800

Tire size: 8.00 x 14

Performance (383 V-8)

Acceleration, 0–60mph: 12–13

Top speed (mph, approx.): 110–120

Fuel mileage (range): 9–15

Production

	1960	1961
Matador	27,908	0
Polara	16,728	14,032

Chapter 13

Fun	Investment	Anguish
3	6	3

Lancer
1961–62

History

A modest car, with much to be modest about? Yes and no. The Lancer was hastily contrived as a sop to Dodge dealers clamoring for a compact, after the successful 1960 introduction of the Plymouth Valiant. Stylists hastily conjured up a full-width horizontal bar grille to distinguish it from its square-grilled cousin. Since Virgil Exner's team had designed the Valiant groundup, harking back to prewar themes with its open wheel wells, sharp-edged fenders and "classic" grille, the Lancer's modern-age grillework looked out of place. And there were some people who thought both of them were homely little toads, grille or no grille.

That's the down side. But by many ordinary standards of judgment the Lancer was a darn good car. Riding a 106.5in wheelbase (the shortest Dodge in history, weighing 700lb less than a Dart), it was of a pleasant, navigable size. The decklid was clean, with absolutely no

Dodge's first "compact," the smallest since its diminutive Wayfarer of the early 1950s, Lancer was a Valiant with a full-width grille and minor trim variations. Shown is a 770 sedan, alongside a much more highly collectible "Clipper" pedal car.

These sports coupes were Lancer's first response to the Corvair Monza, which was popularizing the sporty compact in 1961. A color splash around the beltline and a deluxe, color-coordinator interior of nylon and vinyl distinguished what Dodge called the "Lancer sport coupe."

Lancer was distinctly different from Valiant at the rear, where a clean deck showed no sign of Virgil Exner's "toilet seat" mock spare tire imprint as used on Valiants.

sign of the "toilet seat" spare tire image Ex had implanted on the Valiant. Its interior trim was a definite grade above that of the Valiant. Its lightness, and the standard three-speed manual floorshift, gave it excellent performance, and downright vigorous pickup with the optional 225ci Slant Six.

Lancers came in two flavors: the 170 (named for the standard Slant Six displacement) and 770 (named because it sounded more important than 170). Each offered a four-door sedan and wagon and two-door coupe. A neat little hardtop was also available in the 770 series. Lancer prices started at $1,979, cheaper

The most collectible Lancer is the hardtop, part of the 770 series for 1961, with clean, airy styling and optional two-tone color treatment. Most of these come with the more powerful 225ci Slant Six, one of the few performance sixes of the decade.

Narrow-band whitewalls, amber turn indicators, a new grille, and revised side trim distinguished the 1962 Lancers. The sporty hardtop was now called, with considerable exaggeration, the "Lancer Gran Turismo." It featured all-vinyl, color-keyed upholstery and bucket seats; the 145hp Slant Six was optional, though usually fitted. Lancer GT badge adorns front fenders.

Profile view of 1962 Lancer 770 shows some of its styling problems: an unrelated front and rear end and odd-shaped rear doors. The best thing about this model was the price, scarcely more than $2,100 base.

than Valiant's, and considering that Plymouth had more dealers, Lancer sales of 75,000 against 120,000 Valiants for the '61 model year was encouraging.

Although the standard 225 Slant Six gave fine performance, much more was available from the Hyper-Pak: a hot cam mated with four-barrel carburetor and ram manifold and four-barrel carburetor, which cost $403, and was dealer-installed. Advertised at 196hp, it was said actually to have "in excess of 275hp" by one source. Teamed with a 3.55:1 rear axle ratio and stickshift, *Motor Trend* reported 0–60 times under ten seconds, real V-8 stuff. It's a shame Chrysler never put sufficient performance emphasis on the Slant Six. Derivations of the Hyper-Pak could have powered a new generation of economical but quick sporting cars.

For 1962 the grille, instrument panel and minor pieces of trim were changed and the 770 hardtop was turned into a separate, top-line series, the Lancer GT, tricked out with the carpets, bucket seats and color keyed interior, the sporty doodads that the Corvair's Monza had made popular. Priced under $2,300 and equipped as standard with the 225 engine, this was a good looking, sprightly performer that garnered over 13,000 sales. But it could have been better had Chrysler given it a performance suspension and a four-speed manual gearbox. A folding sunroof, featured on a GT showcar, also didn't make it to production, unfortunately.

In 1963 Chrysler revamped its compacts and Dodge produced the new Dart, now distinctly different in appearance from the Valiant. The Lancer nameplate disappeared, having racked up a commendable 135,000 sales over two model years.

Identification

Valiant-based bodies with horizontal bar grille in 1961 and intricately patterned "vee'd" grille in 1962. Both years featured round taillights (Valiant's were oval). Two-door hardtop model was in the 770 series in 1961 and broken out as the Lancer GT in 1962. Hyper-Pak models will immediately identify themselves either under the hood or off the mark.

Wish they'd built it: the Lancer GT showcar with folding sunroof, metallic paint, special grille, and bucket seat interior. With the 200+ horsepower Hype-Pak this would have made a wonderful top-of-the-line Lancer. Come to think of it, you could build one yourself: the materials are available.

Appraisal

Though the Valiant is a purer statement of Exner's design philosophy, I've always liked the Lancer GT and have kept an eye out for them in the ads, but they are few and far between. Most of the Lancers offered are not GTs or even 770 hardtops, and when those are found they tend to be scruffy. They were never preserved with collectibility in mind. Of course the most desirable are the hardtops, the 1961 770 and 1962 GT, but the upper level 770 coupes are good looking in their way, and certainly had more character than the Falcon or early Corvair; wagons suffer from too many roof pillar angles, however. Thanks to the torsion bar front suspension, Lancer and Valiant outhandle all the other compacts of their day. The neat three-speed manual transmission with floorshift is preferable to Torqueflite. The ultimate Lancer would have the Hyper-Pak, which was usually found in hardtop models but is, unfortunately, rare.

Summary and Prospects

Appreciation for the Lancer has been remarkably good for a six-cylinder car that was never available with a softtop, certainly better than you can get by putting the same money in the bank. Of course in 1982 they were still depreciating, so the baseline for our calculations is the very bottom of the depreciation curve. I don' believe they'll continue to appreciate at the 1982–95 level. Another problem is that 95+ point Lancers are extremely rare. You can buy a condition 3 GT for a couple grand, but it'll cost you far more than $5,000 to put it in show-winning condition, so top dollar for the scarce fine example is well spent. Lancer prices parallel those of comparable Valiants except for the GT, which in top shape is worth $2,000 more than its counterpart, the Valiant Signet.

Price History

95+ point condition 1	1982	1988	1995	Return
sedans	$1,200	3,400	5,000	12.6%
wagons	1,000	3,200	5,000	14.3%
coupes	1,500	3,700	6,000	12.2%
1961 770 hardtop	1,700	4,000	6,400	11.6%
1962 GT	1,800	4,500	7,000	11.9%
(Add 20% for Hyper-Pak)				

Production

	1961	1962
4dr sedans	44,864	28,793
coupes	12,637	14,333
wagons	9,720	7,005
770 hardtop	7,552	0
GT hardtop	0	14,140

Note: Dodge does not distinguish between models for production by body style; however, the proportion of 770s to 170s was a little over 2:1 in 1961 and about 1.5:1 in 1962. From this we can estimate, for example, 8,000 770 coupes against 4,000 170 coupes in 1961, 4,200 770 wagons against 2,800 170 wagons in 1962, and so on.

Specifications

Engine

Type: aluminum block ohv slant six

Six: 170ci (3.40 x 3.13), 101hp standard, all except 1962 GT

225ci (3.40 x 4.13), 145hp optional; standard, 1962 GT; Hyper-Pak special order option, 196hp

Chassis and Drivetrain

Unibody construction.

Transmission: three-speed manual, TorqueFlite automatic optional

Suspension: torsion bar front suspension, solid rear axle with leaf springs

Measurements

Wheelbase (in): 106.5

Curb Weight (lbs): wagons 2,800; other models 2,600

Tire size: 6.50 x 13

Performance

Acceleration, 0–60mph: 170 six, 16; 225 six, 13; Hyper-Pak, 9.6

Top speed (mph, approx.): 95–100; Hyper-Pak, 110

Fuel mileage (range): 18–22; Hyper-Pak, 12–18

Chapter 14

Fun	Investment	Anguish
3	5	5

880-Polara-Monaco
1962–66

History

Unlike Plymouth, Dodge did not pin all its 1962 hopes on a dramatically downsized line of intermediate-size Darts and Polaras. In addition, the Division copped the smooth new bodyshell of the Chrysler Newport, along with its 122in wheelbase, to replace the standard-size Polaras of 1961. Since Dodge held ninth place in production for 1962, while Plymouth fell from fourth to eighth, it might be thought that retaining a full-size model had saved Dodge from disaster. In fact, however, the Custom 880, Dodge's Newport clone, accounted for fewer than 18,000 sales in 1962, and Dodge would have still finished ninth without it.

The Custom 880 remained a peripheral model through its final appearance in 1965, never achieving 10 percent of production, while steadily increasing its sales (as did Dodge as a whole). For the '65 model year it made close to 45,000 sales and a handsome profit. With the demise of the old division structure, which had teamed Plymouth with each Chrysler, DeSoto, and Dodge dealer, Chrysler Corporation evolved at this time into a two-division company comprising Dodge and Chrysler-Plymouth, DeSoto having vanished after 1961. Dodge dealers therefore needed a full-size model, got it in the 1962–65 Custom 880, and retained it in the later sixties with new generations of Polaras and Monacos on the same generous wheelbase.

The 880 was, furthermore, a complete line of cars, with two- and four-door hardtops, a sedan and wagon, even a convertible coupe. Prices in 1962 began at just under $3,000 and were identical to those of the Chrysler Newport—much to the disappointment of Dodge dealers, who thought their version should cost less than a Chrysler. Faced with a choice between almost identical cars bearing a Chrysler or Dodge badge, more customers bought the Newport, which outsold the Custom 880 by better than five to one in their first year of direct competition. Gradually Dodge reaped a better proportion of sales, but never rivaled the Newport in popularity. What this means for collectors is that a Custom 880, car for car, is rarer than a Chrysler Newport. However, the 880 has yet to exhibit higher collector value than the same body style Newport: it's a wash.

Dodge floated a price-leading ($2,815 minimum) trio of detrimmed models called 880s without the "Custom" for 1963–64. In 1965 these base 880s were replaced by a full line of upsized Polaras, which shared the Custom 880's 121in wheelbase. Both models had new sheetmetal, based on that of the larger Plymouths. (The once-cloned Chrysler Newport had added length and moved farther upmarket.) Starting prices for many body styles were still comfortably under $3,000, and between them the Custom 880 and Polara accounted for 120,000 sales.

The 1962 Dodge Custom 880 was a Chrysler Newport with a '61 Dart grille, a stop-gap product planner's creation built to plug a gap in the corporate line-up created by the departure of DeSoto. The marriage worked reasonably well; in fact, many think the '61 grille better suited to the Chrysler body than the 1961 Dodge. Expect no investment value in one of these, however.

Glitzy dashboard and a broad bench front seat were hallmarks of the Custom 880. Except for the nylon seat inserts, interiors of these cars hold up pretty well.

The first Dodge Monaco arrived in 1965 as a premium limited edition of the Polara two-door hardtop, beautifully styled, fitted with a 383 V-8, foam seats, clock, remote control rearview mirror, center console, padded dash, three-spoke steering wheel, and special spinner wheelcovers. At $3,308 base, the Monaco was popular enough to remain in the line for 1966. By then, however, it was called the "Monaco 500," to distinguish it from a full line of Monacos that had replaced the Custom 880, wearing the same basic sheet metal and still on the 121in wheelbase.

The Custom 880 and its Polara/Monaco descendants were typical sixties big cars, powered by 361ci or larger V-8s, tighter than their competition thanks to Unibody construction, and better handlers than most cars their size thanks to torsion bar front suspension. They were, however, plagued by indifferent quality standards.

Identification

Finless Newport body but with full-width Polara-style concave grille encompassing horizontal quad headlamps instead of the Newport's smaller oval grille and slanted vertical quad headlamps. For 1963, a shorter grille opening, still full-width encompassing headlamps, and a stand-up hood ornament. The '64

The 1963 Custom 880 had a more conventional front end, distinctively Dodge, but the rest of the car was pure Chrysler, as in 1962. Four-door hardtops are worth little more than sedans on the collector market.

The only Custom 880 worth a collector's glance is the convertible, a clean piece of styling and rare (only 822 built), but hardly striking from a styling viewpoint. I'd have a hard time getting excited about one of these, unless it was showroom fresh with 10,000 miles on the clock.

For 1965, Monaco (top) was now an entirely new, upsized line of prestige sports models, while the Custom 880 (below) continued as a Chrysler clone with a Dodge front end. The 880 four-door was a six-window design, providing more glass area than any previous Dodge sedan. Both cars rode a 121in wheelbase.

was similar but the grillework was now in two rows, each slanting inward toward the middle, quite similar to the concurrent '63 Rambler Classics. For 1964 the Custom 880, joined by a detrimmed version badged as a Polara, carried senior Plymouth sheetmetal with its distinctive razor-edge side moldings and crisp rooflines. The grillework continued with vertical blades, the grille opening flared out at the edges to encompass the headlamps. Custom 880 four-door sedans carried fixed quarter windows aft of the rear doors, while Polaras were conventional with closed rear quarters. Monacos (1965) came as two-door hardtops only and carried the model nameplate. In 1966 the Custom 880 line was replaced by the full line of Monacos, while the limited edition sports coupe became the Monaco 500. While '65s and '66s are very similar, there is an instant identification: on '65s the ridge over front wheel arches continues toward the front of the car at axle level, whereas on '66s the line continues forward at the height of the tire tops, ending just above the front bumper. Both '65s and '66s feature wide, "delta" taillamps.

Appraisal

These are all V-8s, thirsty cars with a marked preference for the kind of gasoline that is virtually unobtainable today and a tendency to suffer drastically from the dreaded tinworm. The most economical versions are the 1966s with 318 V-8s. On the highway they cruise happily at modern speeds, but are not known for their qualities or fit and finish any more than for their longevity. It pays to buy the rust-free western cars, particularly with low mileage. Big engined cars with the Ramcharger, Max-Wedge, or Hemi engines were theoretically available, but are very scarce. Hemis are highly sought after.

Summary and Prospects

Custom 880s and the 1965 Polara (which is indistinguishable as to value, body style for body style) have returned very consistently and well on investment, assuming one invested in a prime-condition show-quality exam-

Dodge's all-new enlarged Polara debuted for 1965, with a distinctive grille of "dumbell" shape, stand-up hood ornament, novel "C" pillar and deluxe interiors. Like the new '65 Monaco, it rode a 121in wheelbase, and was the largest Polara yet built.

ple. I tend to think it hasn't been the same for run-of-the-mill "condition 3" cars, which are worth little more now than they were ten years ago. Whether these return figures will hold up in the future is a good question. The cars seem to lack something, that indefinable character that makes one vehicle truly collectible and another an also-ran. This writer's opinion is that it's not worth investing in anything less than a "condition 1" example, preferably the convertible or two-door hardtop. (Notice especially the convertible production figures: these are very rare Dodges!)

The "sleeper," if there is one, is the 1965 Monaco, first appearance of this now-famous Dodge name, sold only as a razor-edge two-door hardtop and carrying a 315hp version of the solid 383 V-8. Its current price range is no higher than that of more standard 880 hardtops, which makes it a good buy, but you first have to find one.

Price History (Non-Hemis)

95+ point condition 1	1982	1988	1995	Return
convertibles	$2,600	6,500	12,000	13.5%
2dr hardtops	1,500	4,700	7,000	13.6%
4dr hardtops	1,250	4,000	5,500	13.1%
other models	1,000	3,500	5,000	12.2%
1965–66 Monaco/Monaco 500	2,000	5,500	7,000	11.0%

Production

	1962	1963	1964	1965	1966
Custom 880 4dr sedan	11,141	9,233	9,309	9,380	0
Custom 880 4dr hardtop	1,855	2,564	3,634	7,966	0
Custom 880 2dr hardtop	1,761	2,804	3,798	4,850	0
Custom 880 6 pass. wagon	1,174	1,647	1,639	4,499	0
Custom 880 9 pass. wagon	890	1,365	1,796	5,923	0
Custom 880 convertible	684	822	1,058	1,416	0
880 4dr sedan	0	7,197	7,536	0	0
880 6 pass. wagon	0	1,727	1,908	0	0
880 9 pass. wagon	0	907	1,082	0	0
Polara (121in wb) 4dr sedan				*75,000	*75,400
Monaco/Monaco 500				13,096	7,300*
(121in wb) 2dr hardtop					
Monaco					30,600*

*approximate

Specifications

Engines

Types: cast iron ohv V-8
318ci (3.91 x 3.31), 230hp, standard on 1966 "Polara 318"
361ci (4.12 x 3.38), 265hp, standard 1962–64
383ci (4.25 x 3.38), 305 & 330hp, optional 1963–64; 270hp, standard 1965–66 optional Monaco 500 1966; 315 standard Monaco 1965; 325hp, standard Monaco 500, optional on others1966
413ci (4.25 x 3.75), 410–420hp, optional 1962; 340–90hp, optional 1963; 365–425, optional 1964–65; 340, optional1965
426ci (4.25 x 3.75), 415 & 425hp, optional 1963
426ci Hemi (4.25 x 3.75), 400–425hp, optional 1964; 415–425hp, optional 1965
440ci (4.32 x 3.75), 350hp, optional 1966

Chassis and Drivetrain

Unibody construction.
Transmission: three-speed manual, four-speed manual (1965–66), TorqueFlite automatic
Suspension: torsion bar front suspension, solid rear axle with leaf springs

Measurements

Wheelbase (in): 122 in 1962–64, 121 in 1965–66
Curb Weight (lbs., average): wagons 4,250, convertibles, 3,900, others 3,700.
Tire size: 8.00 x 14 in 1962–64, 8.25 x 14 in 1965–66, .55 x 14 in wagons, 1965–66

Performance (383/413 V-8s)

Acceleration, 0–60mph: 9–11
Top speed (mph, approx.): 110–125
Fuel mileage (range): 10-15

Fun	Investment	Anguish
7 (V-8)	6	4

Little Darts
1963–66

History

For 1963 Dodge dropped its Valiant-clone, the Lancer, and introduced the Dart: an old name on a new package. The compact Dart was destined to be one of the most successful compact cars in history. It lasted fourteen years, virtually the same size and wheelbase for all of that time, accounting for over 2,750,000 sales, and in many of those years it was one of the few success stories Chrysler Corporation could claim. When it was finally replaced by the Aspen in 1977 (a dubious successor which crash-dived and was shortlived) the Dart was genuinely mourned and missed.

Markedly larger than the new '63 Valiant, the Dart immediately established its own identity. The last traces of Virgil Exner's long styling reign had now vanished, and the new

Darts shared the production line with Dodge intermediates in 1963. Loosely based on, but quite different from the new '63 Valiant, the Dart proved a durable product, and ably maintained Dodge's volume in some of the thin years ahead. It lasted fourteen years and racked up nearly three million sales: nobody can sniff at that.

What the Lancer had desperately needed, its replacement had from the beginning: a convertible. Properly equipped and restored from the demon rust, these are still satisfying little cars. If you have a choice, the GT convertible is preferable to the 270 model.

body had been styled by Elwood Engel, late of Ford, who gave it crisp, clean lines with a sharp-edged extruded look through mid-body. Compared to the Valiant's rather dowdy lines, the Dart looked clean, sprightly and ready for action—yet it retained a family resemblance to its larger Dodge cousins. Following previous Lancer practice, there were two trim levels plus a more limited production sporty confection called "GT." A true gran turismo this was not—but it had the colorful interiors, bucket seats, optional stickshift, and exterior styling distinctions made popular by the Corvair Monza/Falcon Futura a few years earlier. In 1963, Dart's standard and only engine was the Slant Six, but as before, customers could specify the big-inch version with its energetic 145hp. But the potent Valiant 273 small-block V-8 became optional in 1964, making for a far more exciting package.

Dart prices started at under $2,000, the cheaper of two convertibles at $2,385, and a loaded GT ragtop didn't run much more than $3,000. With good looks, decent performance and economy, it was a formula that couldn't miss. Sales took off, and by 1965 Dart was outselling Valiant.

Identification

These Darts are easily identifiable: all years feature dual headlamps rather than quads; fenders jut forward at the front and backward at the rear; rear windows on two-door models have a neat "notch" in the style of BMWs (though BMWs they're not). The concave 1963 grille carries the busy vertical pattern of the Dodge family that year and the "Dodge" name is on the hood. For 1964 the grille was cleaned up with "Dodge" on a cen-

Clean wagon styling distinguished the 1963 Dart wagons, this being a low-priced 170 model. Wagons, as usual, have no collector value.

A minor facelift with a broad grille nameplate distinguished the 1964 Dart line. This is a GT hardtop, distinguished by its bucket-seat interior and badge on the roof quarter. These are cheap wheels on the collector market but beware the dreaded tinworm.

tral bar. For 1965 the nameplate was in script on the right of the hood while the grille was composed of heavy horizontal and vertical bars; and in 1965 "Dodge" went back across the hood in block letters while the grille was made up of three sections of fine horizontal bars. Hoods of 1962–63 models carry dummy hood scoops.

Appraisal

Darts with the energetic 273 V-8 are great fun to drive, especially if you luck onto one with the stickshift, preferably the four-speed. Rated at 180hp (gross) by the factory, it was capable of being tuned to deliver much more. Like the Slant Six, this was a reliable powerplant built to last a long time, and parts for both engines are in good supply as this is written. Body parts are much scarcer, and the amount of lightweight aluminum brightwork, which is easily damaged, can make replacement a nightmare. Still, complete restorations are at least conceivable on Darts; they will probably cost more than the finished product is worth, but if you find the model you like, you may consider it money well spent. V-8s are especially worth seeking out because they are relatively rare, especially in the 170 model—although this was considerably detrimmed and ranks mainly as a stop-light Q-ship—or in their "high performance" state, with four-barrel carburetor and 235hp (1965–66). The post-1963 GT offers the best likelihood of a V-8 under the hood, and in this form is the most desirable version.

Summary and Prospects

Early Dart compacts have done very well indeed during the recent recession, in which it is imagined that virtually all collector cars took heavy plunges in value. Darts have the advantage of being eminently practical for today's conditions, able to get by on the lean stuff we call modern gasoline, while returning downright fine performance from the V-8 and good performance from 225 Slant Six. I don't think the collector crowd has really discovered them yet: strongly recommended for future appreciation.

Dart GT hardtop for 1965 wore a new roof trim and had a new hood, deck, grille, bumpers, and taillights. From mid-1964, Darts had been available with the smallblock V-8, and this is definitely the engine to look for on '65 and later models.

Another GT convertible, this one from 1966, with ultra-clean styling: well worth a glance if you find one in nice condition. Four-speed manual transmission and smallblock V-8 make a sprightly combination on these good-looking cars.

Price History

95+ point condition 1	1982	1988	1995	Return
GT convertibles	$2,250	5,500	9,000	12.2%
270 convertibles	2,000	5,250	7,000	11.9%
GT 2dr hardtops	1,500	5,000	8,000	14.9%
270 2dr hardtops	1,300	4,500	6,000	13.5%
other models	1,000	3,250	4,000	12.2%
Add 10% for V-8s				

Production

	1963	1964	*1965	*1966
170 (6 cyl)	58,536	74,625	70,900	28,400
170 (V-8)	0	2,509	2,900	1,400
270 (6 cyl)	61,159	58,972	52,900	28,500
270 (V-8)	0	7,097	9,900	6,600
GT (6 cyl)	34,227	37,660	22,700	8,700
GT (V-8)	0	12,170	18,000	10,000

* approximations

Specifications

Engines

Types: aluminum block ohv slant six & cast iron ohv V-8
Sixes: 170ci (3.40 x 3.13), 101hp standard
225ci (3.40 x 4.13), 145hp optional
V-8s: 273ci (3.63 x 3.31), 180hp optional 1964–66; 235hp optional 1965–66

Chassis and Drivetrain

Unibody construction
Transmission: three- and four-speed manual, TorqueFlite automatic
Suspension: torsion bar front suspension, solid rear axle with leaf springs

Measurements

Wheelbase (in): 111, wagons 106
Curb Weight (lbs): 2,600–2,800
Tire size: 6.50 x 13

Performance (235hp V-8s)

Acceleration, 0–60mph: 9
Top speed (mph, approx.): 110
Fuel mileage (range): 12–18

Chapter 16

Mid-Size Dodges

1963–70

Fun	Investment	Anguish
	1963–64 330, 440, and Polara	
4	4	6
	1965–70 Coronet	
4	4	4
	500, R/T, and Super Bee:	
8	7	7
	Ramcharger/Hemicharger:	
10	10	10

History

It's hard to categorize the size of these cars, since they were bigger than the ordinary "intermediate" yet smaller than what was then considered "full-size." They represented another Dodge attempt to garner harvests in fields unplanted by rivals from GM, Ford, and even Chrysler-Plymouth. Over the years the Coronet family shrank slightly, and by 1970 they were genuine intermediates. With some of the most likable and exciting Dodges ever built, these cars crystallized the Dodge Boys image of high performance which persists to this day. In more mundane permutations, the Coronet and its midsize predecessors were the meat and potatoes of the Dodge line-up in the sixties. Along with the compact Dart, Coronet sales were an annual indicator of Dodge's success.

Fully restyled and fin-free, the 1963–64 midsize Dodges comprised three models with increasingly superior trim: 330, 440, and Polara. Each was available with the 225 Slant Six or a 318 V-8, although the body mix was different depending on the market target (no 330 hardtops; convertibles only in the Polara line and only with V-8s). The four-door hardtop body was temporarily eliminated. All intermediates rode 119in wheelbases except wagons (116). For Polara convertibles and hardtops, a 500 trim package created the ultimate sporty Dodge, with bucket seats, console, and a 330hp version of the reliable 383ci V-8. Top eliminator for the dragstrips that year was a mighty Ramcharger 426 with 415hp or 425hp. The latter, called "Max Wedge," had dual four-barrel carbs, pop-top aluminum pistons and high lift cam; it creamed a passel of National Hot Rod Association records. The model line-up was repeated in 1964, Dodge's fiftieth anniversary year, with a neat front end facelift and new, V-shaped "C" pillars for two-door hardtops. The big news this year was the reintroduction of the Hemi: hemi-segment heads on Ramcharger engines. "Hemichargers" were rated at 400–425hp but probably developed well over 500, and supercharged Hemis in a pair of light 330 two-doors turned quarter-mile elapsed times at 135mph in exhibitions.

In 1965 Dodge brought the old Coronet name out of retirement to designate its entire line of genuine intermediates in three levels of trim, base, 440, and 500, while the Polara name moved up to the senior models (see Chapter 14). A few Hemis were now sold for street use, although Chrysler disassociated itself from any responsibility and voided their warranties. At the other end of the scale, six-cylinder convertibles were available, though rarely ordered.

Reskinned with a sharp-cornered body pinched in the middle, the 1966 Coronet was one of the best looking Dodges in history. The following year Dodge extended its range with

Representing a new line of clean-limbed, fin-free standard size Dodges, the 1964 Polara 500 poses at the World's Fair. Most interesting was its "V-type" rear roof pillar, common to Plymouth as well as Dodge, a late innovation by the departed Exner.

A Polara four-door hardtop from 1963 displayed the peculiar slotted grille of that year. The only standard '63s worth pausing over are convertibles; closed models should not be considered unless in exceptional low-mileage original condition.

a leading-edge performance version called Coronet R/T (for "Road/Track"), available as a convertible or hardtop, with a standard 440 V-8, bucket seats and luxury interior. R/T hardtops could be had for around $3,500 fully equipped; convertibles generally cost around $4,000. They were terrific cars for the money and are still sought after with all the enthusiasm of 1967. At the time, in the height of the musclecar era, R/Ts ran with the best. For 1968, which was a repeat of the '67 model lineup with a slight trim shuffle, R/Ts were part of the Dodge "Scat Pack," identified by double vertical "bumblebee" tail stripes.

A clean-limbed 1964 Polara 500 convertible, with bucket seats, console, padded dash, deluxe wheel covers, and console shifter, this is one of the nicest looking Dodges of the 1960s. These models can still be had in prime condition for well under five figures, and strike me as quite a bargain. Now all you have to do is find one. . . .

An old name brought back for 1965 was Coronet, which had some interesting years ahead, what with Hemi and 440 engine options. You're not likely to find one of these, but owning one would be fun. Police specials were equipped by Dodge with big-block engines and heavy-duty suspension systems, and were among the most popular cruisers of the day.

A competitive new model for 1969 was the Coronet Super Bee hardtop and coupe, counterpart to Plymouth's Road Runner, based-priced at just over $3,000, and equipped with a 335hp version of the 383 V-8. Coronets retained their clean-limbed body styling; useful innovations were power front disc brakes and dual braking with two master cylinders.

This generation of Coronets came to an end in 1970, when a curious "dual-loop" arrangement of divided bumper/grilles was combined with huge delta-shaped taillights; the model line-up continued as in 1969, with the same arrangement of body styles.

Identification

A big, forward-jutting, vertical bar grille with small inboard headlights announces the 1963 models, while the '64s line up the headlamps side by side with a shorter, neater grille and higher front bumper. Four-door hardtops: none in 1963, Polara only in 1964.

Starting in 1965 all models were called Coronets, which featured a bland grille of thin vertical bars, the Dodge name in broadly spaced letters on the hood, and the Coronet badge on front fenders. A divided, thin-mesh grille distinguished 1966 models, which were completely restyled with sharp corners, a pinched waistline and plentiful body side creases. The '67 was facelifted, returning to the thin vertical-bar grille and a central medallion. For 1968, blacked out grilles with Dodge letters and dummy reverse scoops just ahead of the rear wheel cutouts were distinguishing features. The 1969 Coronets received a thin grille shell, flared at outer ends, blacked out and divided on 500s, R/Ts and Super Bees. A divided, flared "double loop" bumper grille distinguished the 1970 models.

Appraisal

I think the 1966–69 Coronets, particularly the two-door hardtops, are among the best looking American cars of their day. Combined with a great range of performance engines (and it doesn't take a Hemi to make a hot Coronet), excellent roadability and the possibility of four-speed gearboxes, they offer a lot of attractive features. On the other hand, they were not well assembled, and the lack of quality control thirty years ago means they haven't held up as well as they might have otherwise.

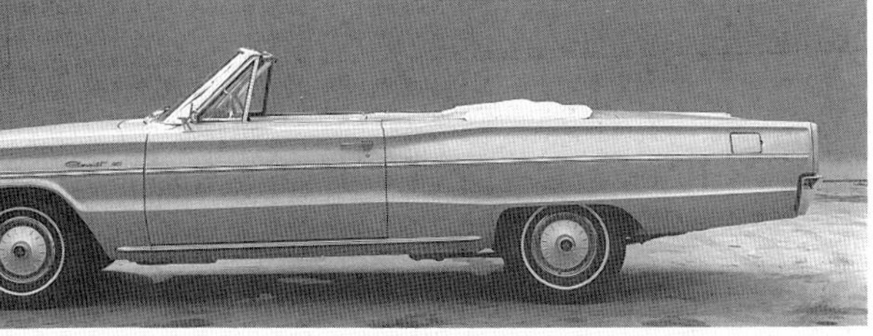

Handsomely styled with a hint of Coke-bottle to the side profile, the 1966 Coronet 440 convertible is highly desirable and still likely to be found in fine condition for under $10,000. These cars did occasionally come with six-cylinder engines, but stick with the V-8s for maximum investment value.

If you can't afford a convertible there's always the two-door hardtop, which looks especially nice with its novel rear roof quarter treatment. Styling for this car was directed by Elwood Engel, late of Lincoln, who brought some of his better ideas to Dodge.

Higher performance versions of the 383 and all larger V-8s really require fuel that isn't readily available anymore, and that poses driveability problems. For all around collector use, everyday driving as well as the occasional drag race, the 383 and smaller V-8 is a better bet than the 400-plus inchers; but they don't pack the sheer power, mind-blowing acceleration and crowd pleasing excitement of the big-blocks. So this is a personal decision. By the way, *all* Coronet sixes had the 145hp, 225ci Slant Six, a fine powerplant offering decent performance, especially when teamed with a manual transmission. Lighter models equipped with this engine have only around 20lb per horsepower, a pretty good ratio for a six. Engine parts are relatively easy to find for most Coronets, but body parts and especially exterior trim are hard to come by. Finally, watch out for the demon rust, which plays hell with these unit body cars, as it does all Chrysler products of their kind.

Summary and Prospects

The Super Bee was Dodge's version of the Road Runner, not as well known and scarcer, but just as quick, standard with a 335hp 383, optional with the Hemi. The 440 Six Pack was available in 1969–70. *Automotive Investor*, "the insider's newsletter for savvy collectors," predicted steadily negative returns on investment for virtually every R/T and Super Bee in their January 1994 "annual look"; yet six months later they were rating the 1969 Super Bee Hemi as a "hold," predicting it would go from $30,000 to $36,000 in five years and to

$45,000 in ten years (condition 1 show quality). That's only 4.5 percent return on your money, compounded annually, and only goes to prove that *Automotive Investor* has no better a crystal ball today than it did when I edited it some years ago.

What we *can* do with every confidence in accuracy is look at the performance of the hotter Coronets over the years and then make a personal judgment as to where they're headed as collector's items in the next decade. (Please note that the Hemis and Six-Packs have actually *depreciated* from the heights of 1988, when people were bidding them out of sight and someone who should have known better paid six figures for a Plymouth Hemi 'Cuda at auction.

The rates of return shown below are misleading in the case of special engine jobs because they've depreciated lately, while still returning well on your investment if you've owned one for twelve years. Nevertheless, you would have enjoyed a much higher and steadier rate of return had you invested in a plug-ordinary 1965 Coronet 440 convertible, or a 1969–70 R/T with the standard-issue engine.

What conclusions can be drawn? The most obvious is that convertibles and two-door hardtops are, in that order, always the best investments; that *sporty* convertibles and hardtops are among the better of the best; and that big-engine versions like Ramchargers, Hemichargers, and Six-Packs are speculative. Since there appears to be no likelihood of the acquisitive fever of the late 1980s returning, auction shills and "Hemmings Hopefuls" will not be bidding up big-block Dodge prices for the foreseeable future. Left to the collector market, they'll probably appreciate gradually in value the way God and the car collector intended.

Ergo: every 1963–70 Dodge midsize convertible and two-door hardtop (including the hardtop-like Super Bee coupe) is a good buy; sedans and wagons aren't, unless they are in exceptional low-mileage mint condition. Four-door hardtops are in between, but none were built in this grouping except the 1964 Polara.

Rarities include all kinds of oddballs. Coronet 500s were generally expected to be V-8s, but Dodge built 500 sixes in 1966 and 400 in 1967—do you really care? Hemis are far more likely to be encountered in Chargers and Challengers than Coronets, as indicated by Dodge's own figures. For example, Hemis were installed in only thirty-two 1970 Super Bee hardtops (twenty-one four-speeds, the rest

The Super Bee Coronet joined the 1968 Scat Pack family of sporty cars including the Coronet and Charger R/Ts and Dart GT Sport. Super Bees came with a 383ci, 335hp engine, and the Hemi was optional. Hardtop roof styling, simulated hood scoop, wide tires (redwall or whitewall) and bumblebee racing strips distinguish this Dodge counterpart to Plymouth's Roadrunner.

automatic), and only one R/T convertible (four-speed). You can imagine what that ragtop is worth.

One other thing: 1970 marked the last Dodge convertible for many years, not only in the Coronet but in the Polara line, and (a word to the convinced) production was minuscule, making '70 ragtops very desirable.

Price History

95+ point condition 1	1982	1988	1995	Return
1963–64 330 and 440 2dr hdtp.	$1,300	4,200	6,000	13.5%
1963–64 Polara 500 conv.	2,400	6,500	7,500	9.9%
1965 440 convertible	1,500	6,500	9,500	16.6%
1965 500 hardtop	1,800	5,500	9,000	14.3%
1965 500 convertible	2,500	7,000	11,000	13.1%
1967–68 R/T hardtop*	5,000	8,500	10,000	5.9%
as above, Hemi engine	7,500	40,000	30,000	12.2%
1967–68 R/T convertible*	5,750	9,500	15,000	8.3%
1968–69 Super Bee*	3,300	9,500	10,000	9.6%
as above, Hemi engine	7,000	40,000	30,000	12.2%
as above, 440 Six-Pack engine	3,500	25,000	18,000	14.6%
1969–70 R/T hardtop*	2,400	10,000	27,500	22.4%
1969–70 R/T convertible*	3,200	11,000	30,000	20.4%

* Prices are for standard-equipment engines. Add 100%+ for Hemi, 50% for 426 wedge or 440 magnum

Specifications

Engines

Types: aluminum block ohv slant six & cast iron ohv V-8
Six: 225ci (3.40 x 4.13), 145hp standard 1963–70
V-8s: 273ci (3.63 x 3.31), 180hp standard 1965–67; 190hp standard 1968–70; 235hp optional 1965
318ci (3.91 x 3.31), 230hp standard 1963–64, optional 1965–70
361ci (4.12 x 3.38), 265hp optional 1965–66
383ci (4.25 x 3.38), 270hp optional 1967; 290hp optional 1968–70; 305hp standard 1963–64 Polara 500, optional 1963–64; 315hp optional 1965; 325hp optional 1966–67; 330hp optional 1963–65, 1968–70; 335hp standard on Super Bee 1968–70
426ci (4.25 x 3.75), 365hp optional 1964–65
Ramcharger 426ci (4.25 x 3.75), 415/425hp optional 1963–64
Hemi 426ci (4.25 x 3.75), 400/415/425hp optional 1964 (usually competition only); 415/425hp optional 1965; 425/500 optional 1966; 425hp optional 1967–70
440ci (4.32 x 3.75), 365hp optional 1966; 375hp (Magnum) optional 1967–70; 390hp (Six-Pack) optional 1970

Chassis and Drivetrain

Unibody construction
Transmission: three-speed and four-speed manual, Torque Flite automatic optional
Suspension: torsion bar front suspension, solid rear axle with leaf springs

Measurements

Wheelbase (in): 1963–64: 119, wagons 116; 1965: 117, wagons 116;
1966–70: 117
Curb Weight (lbs): 3,000–3,500
Tire size:
1963–64: 7.00 x 14, wagons 7.50 x 14; 1965–66 7.35 x 14, wagons 7.75 x 14; 1967–70: 7.35 x 14, R/T 7.75 (F70) x 14, wagons 8.25 x 14

Performance

The Ramcharger 426 wedge of 1963–64 has been eclipsed in reputation by the Hemi, but was a tremendous powerplant in its own right. A rail dragster equipped with this engine set a new NHRA AA/D record of 190.26mph at a Connecticut dragstrip in 1964. Ramchargers and Hemis turned 0–60 in under seven seconds, Six-Packs in close to that, standard V-8s in 9–11, sixes in 16. In a 1967 match test, Motor Trend worked a Coronet Hemi down to 6.2 seconds in the 0–60 leap and a 440 Magnum (375hp) down to 6.5. Quarter miles were worked down to 13.6 seconds at 103mph for the Hemi against 14.7 seconds and 96mph for the Magnum. Top speed varied with axle ratios; properly geared, light-bodied Hemis could approach 175mph. Remarkably, even driven hard, a Hemi would make 9–12mpg, the Magnum about 1 mpg more.

Production (nearest 100)

	1963	1964	1965
330* six non-wagons	40,100	44,800	37,100
330* V-8 non-wagons	24,000	31,600	26,000
440 six non-wagons	10,000	10,200	11,900
440 V-8 non-wagons	34,300	48,500	75,600
330*/440 six wagons	5,400	5,700	3,100
330*/440 V-8 wagons	20,700	24,600	21,800
Polara* sixes	2,200	2,200	
Polara* V-8s	37,600		62,700
Polara 500 V-8s	7,300	18,400	
Coronet 500 V-8s			33,000

*in 1965, 330 became "Coronet" and Polara was full-size (see Chapter 14)

	1966	1967	1968	1969
(non-wagons only, 1966–68)				
Coronet six	7,700	0	0	12,500
Coronet V-8	3,000	0	0	17,900
Coronet Super Bee V-8	0	0	0	27,800
Coronet Deluxe six	5,600	14,100	19,300	0
Coronet Deluxe V-8	20,600	13,700	25,700	0
Coronet 440 six	14,000	8,600	8,200	4,700
Coronet 440 V-8	96,600	83,900	95,300	101,200
Coronet 500 six	500	400	0	0
Coronet 500 V-8	52,200	28,900	30,100	32,100
Coronet R/T hardtops	0	9,553	*10,400	*6,800
Coronet R/T convertibles	0	628	*500	*400
wagons (broken out separately, 1966–68 only):				
six	3,100	3,300	2,600	
V-8	24,600	20,900	30,500	

* estimates

1970 Production

	sedans	coupes	wagons	hardtops	convs.
Coronet	7,894	2,978	3,694	0	0
Super Bee	0	3,966	0	11,540	0
Coronet 440	33,528	1,236	7,736	24,341	0
Coronet 500	2,890	0	5,172	8,247	924
Coronet R/T	0	0	0	2,319	296

Chapter 17

Fun	Investment	Anguish
7	5	6
	Hemi:	
9	9	10

Charger
1966–67

History

Dodge's response to a mid-sixties (and very temporary) craze for fastbacks, the Charger was derived from Coronet running gear with styling cues from the Charger and Charger II show cars. Popular cliches included hidden headlamps, full-width taillights and a fore-to-aft center console which limited seating strictly to four. These were combined with genuinely useful features like bucket seats and a pleasing fastback shape with fold-down rear seats that created a huge cargo area. The Charger did not look as hastily contrived as such rivals as the Rambler

The early Chargers were part of a shortlived genre of fastbacks created from standard production hardtops. Among these, it was a design standout, with a distinctive grille, hidden headlamps, and clean extruded body sides. A wide variety of power was available, all of it V-8, from 318 to 426ci Hemi.

New wheels, front fender turn signal indicators identify the 1967 Charger; '67s also offered a vinyl roof, which frankly looks awful on this fastback body style. This one sports a 426 Hemi badge, which would make it highly desirable in a modern day auction.

Marlin. Charger's standard engine was the long-established and reliable Chrysler 318 V-8 with 230hp, but options ran all the way to the 426 Hemi were available in 1966, and to the 440 in 1967.

Dodge's satisfaction with the Charger package—and it was very well received by testers and the public—was shortlived because after an initial spurt it didn't sell. Against nearly 40,000 of the '66 model dealers barely got through 15,000 of the '67. Deciding to change its shape from that of a distinctive fastback to a more conventional hardtop in '68 wasn't difficult, and took place fairly late, during the latter part of 1966.

Identification

Early Chargers are easily identified by their fastback shape and disappearing headlamps, the latter one of few such arrangements that look equally good open or closed, thanks to thin vertical blade grillework with invisible seams separating the headlamp units. Styling was nearly identical in both years; 1966 models carry dummy knock-off wheels while 1967s are five-spoke mag-types; the '67s also feature fender-mounted turn signal indicators.

Appraisal

The Charger is a fun-to-drive, biggish sporty car whose characteristics are typical mid-1960s: plenty of power, not enough brakes, fairly slipshod assembly quality, fairly deft styling. A factor possibly affecting your choice of power team is that the four-speed manual gearbox was not available with the 318 engine. Another useful item you might want to look for is the Rallye suspension (stiffer torsion bars, rear springs and shocks) which improves handling considerably. One negative factor you can't get away from: visibility through the rear window is terrible: either the rearview mirror is too low, or the top of the backlight is too high. Expect considerable rust and a devilish time finding body hardware; presence of vinyl upholstery makes interior work relatively easy.

Summary and Prospects

If you look back only as far as the late eighties, you could be a pessimist and say the big-block Chargers have fallen off considerably. But if you take the longer view, allowing for the blip of late eighties auction-driven action, you can see that an early Hemi Charger was a perfectly good although not outstanding investment. If you bought a show-condition example in 1982, for instance, it would have probably set you back no more than $500 or so compared to an ordinary 318. (The Charger 440 has held its value since 1988, but returned almost as well on your money over the 1982–95 period.)

With the four-speed and a performance engine, the Charger is as good a buy today as it ever was, and the very scarcity of Hemis makes these a special bargain at mid-nineties price ranges. Although such authorities as *Automotive Investor* think Hemis will continue to reflect negative returns in the short haul, everyone agrees that in the long term the trend is up. Remember, however, that as a closed car, the 1966–67 Charger will never gallop in value like a convertible during market spurts; also bear in mind that the big mid-1960s fastbacks have never been hot items on the collector market.

Price History

95+ point condition 1	1982	1988	1995	Return
318/383 V-8s$	5,800	9,000	14,000	7.6%
Hemi V-8s	6,500	35,000	28,000	12.9%
440 V-8s (1967 only)	6,000	20,000	20,000	10.5%

Production

1966	1967
7,300	15,788

Specifications

Engines

Types: cast iron ohv V-8
318ci (3.91 x 3.31), 230hp standard; 260hp optional.
361ci (4.12 x 3.38), 265hp, optional 1966
38ci (4.25 x 3.38), 325hp, optional 1966
426ci (4.25 x 3.75), 425hp, optional

Chassis and Drivetrain

Unibody construction
Transmission: four-speed manual (not available on 318) and TorqueFlite automatic
Suspension: torsion bar front suspension, solid rear axle with leaf springs

Measurements

Wheelbase (in): 117
Curb Weight (lbs): 3,500
Tire size: 7.35 x 14 standard, 7.75 x 14 optional

Performance

(383 V-8 with TorqueFlite)
Acceleration, 0–60mph: 9
Top speed (mph, approx.): 110
Fuel mileage (range): 12–17

Fun	Investment	Anguish
5	3	5

Full-size Dodges
1967–77

History

There is precious little collectible among the big Dodge lines from this period, and collector interest drops out of sight after 1970. Nevertheless there are several models deserving of mention among the final generation of mega-wheelbase Dodges, called Polaras or Monacos, through their disappearance in 1977.

Almost the last of the behemoths: the ornate 1975 Royal Monaco Brougham two-door "hardtop" (actually a two-door sedan), complete with opera windows and a Mercury Marquis style grille. This huge Dodge sold for $6,000 as typically equipped; massive crash-absorbing bumper was standard issue in those years.

The last full-size Dodge convertibles from Dodge were the 1967–70 Polaras. For the 1967 –69 period, they were also available in the Polara 500 trim variation, featuring bucket seats and console, which is naturally the preferable version. During the 1970 model year Polaras were sold for the first time with six-cylinder power, and though a Polara 500 remained in the range and included a convertible, the "500" was now just a submodel designation, not a limited edition. Aside from a last-gap run of Challenger convertibles in 1971, the 842 open Polaras for 1970 were the last Dodge ragtops until Lee Iacocca revived the genre on the K-car platform in the mid-1980s. Polaras and Monacos were traditional big American land yachts, approaching two tons with a load of passengers, their standard power coming from the 383 (1967–68) or 318 (1969–70) V-8s and milder versions of the 440 available as an option. (Polaras were never offered with Hemi or 440 Magnum engines.)

The bucket seat Polara 500, last of the full-size sports Dodges in the tradition of the Chrysler 300, was also sold as a hardtop, which most of them were, and this model is also collectible. Further up-market in 1967–68 only was the Monaco 500, the largest and most luxurious Dodge in the line, comprehensively equipped and starting at around $4,100, $500 more than the Polara 500.

Things went downhill fast in the seventies. For 1970 the limited edition "500s" were dropped and the following year there were no Polara convertibles. Starting in 1972, power was expressed in the more honest net horsepower (rather than gross); and after 1973 the Polara name was honorably retired after a fourteen-year run. The surviving Monaco was now the only big Dodge left, a conventional family chariot, mainly sedan or wagon models, and in this form it gradually faded away. By 1977 the Monaco badge had been transferred to an intermediate of no distinction, leaving the big Royal Monaco sedan and hardtop, last of the truly big Dodges.

The nicest Dodge biggies were the sixties models, like this clean-lined Monaco four-door hardtop from 1967, packing a 440 V-8. Pillarless four-doors had their heyday in the late fifties and at Dodge, as at most other manufacturers, they were dropped by the turn of the decade.

Identification 1967–70

Fully restyled for 1967, Polara featured large flared taillights whose delta-shape was repeated in the grille, which had a prominent center section and inset flanking grilles. Monaco's grille was blacked out. Two-door hardtops were semi-fastbacks with reverse slant rear quarters. The '68s were almost identical but can be dated by the model name in block letters on the front fenders (shifted from the rear fenders on 1967s). The full-size body was redone again for 1969, sharing its more rounded, "fuselage styling" with the senior Chryslers and featuring an inset three-hole grille flanked by quad head-

All new, longer, lower and wider than ever, the 1967 standard size Dodges featured "delta" styling, extruded lines, Thunderbird-style roofs and heaps of power. The Monaco 500 stood at the top of the Dodge line, consisting only of a two-door hardtop, with 383 V-8 as standard.

The 1967 Polara convertible, fine examples of which can be had for as little as $5,000 presently, unless it's a Polara 500, which will run 50 percent more. 500s can be distinguished by a small round medallion behind the "Polara" name on the rear fenders.

lamps. For 1970 the hood was wrapped further down at the front, resulting in a short combination bumper/grille.

Appraisal

Everything ever said about big American cars applies to these rolling whales, with one exception: thanks to torsion bar front suspension, they outhandled their Ford and GM competition. With the 383 engines, 1967–68 Polaras were also quite hot, but standard power went to 318 for 1969 and 383 and 440 V-8s were optional. Expect driveability problems on modern gasoline for the 383 and larger engines. Quality of fit and finish was generally not good, and rust remains a severe problem. They are fine highway cars, but shoving one around town is a chore.

Vinyl is easy stuff to restore, but what if your Dodge has an odd pattern cloth upholstery unique unto itself? Happily there's a possible answer in a firm called Legendary Auto Interiors, 121 West Shore Blvd., Newark, NY 14513 (telephone 800-363-8804), who specialize in 1957–80 Chrysler product seat upholstery, door panels, and other soft trim. I mention it here in case I forget it elsewhere; this is a unique service which could solve difficult restoration projects, and no, I don't know the principals or have any interest in the firm.

Summary and Prospects

As long as you stick to the obvious collector cars, convertibles and 500 hardtops, these cars are sound investments, as they have been for the past decade. Post-1970 hardtops are doubtful buys, however, and don't even think about making money with a sedan or wagon. The exception would be the extremely low-mileage, mint original Monaco, particularly if it's a 1977 model (last of the biggies). Such a car is always worth a premium no matter what make or model it is, but don't plan to put a lot of mileage on it because then it reverts to being just one of the pack.

This 1970 Polara hardtop features the controversial "Super-Lite," a quartz-iodine contraption designed to burn the eyeballs off oncoming drivers if they caught it right (although its intent was just the opposite). The Super-Lite was banned in quite a few states, though they've probably forgotten it by now.

Humongous Dodges trundled through the seventies with little styling change from year to year. This 1971 Monaco can be distinguished from the '70 only by its slightly altered, nondivided grille. The 383 was still standard.

Price History

95+ point condition 1	1982	1988	1995	Return
1967–69 Polara 500 conv.	$2,500	7,000	9,000	11.2%
1967–70 Polara 500 hardtop	1,500	5,000	5,500	11.4%
1970 Polara 500 convertible	2,500	6,000	8,500	10.7%
1967–68 Monaco 500 2dr htp.	1,800	5,200	7,500	12.6%
1971–77 two-door hardtops	2,000	3,500	3,500	4.8%

Production

	1967	1968	1969
Polara	69,798	99,055	83,122
Polara 500	5,606	4,983	5,564
Monaco	35,225	37,412	38,566
Monaco 500 hardtop	5,237	4,568	0

Selected Production

	1970	1971	1972	1973
Polara 500 convertible	842	0	0	0
Polara 500 2dr hardtop	15,243	0	0	0
Monaco 2dr hardtop*	3,522	3,195	7,786	6,133
Polara 2dr hardtop	0	11,500	8,212	6,432
Polara Custom 2dr hardtop	0	9,682	15,0391	7,406
Polara Brougham 2dr hardtop	0	2,024	0	0

* "Monaco 500" in 1970

	1974	1975	1976	1977
Monaco 2dr hardtop	3,347	2,116	0	0
Monaco Custom 2dr htp	10,585	0	0	0
Monaco Royal 2dr htp**	0	4,001	2,916	3,360
Monaco Brougham 2dr htp.	4,863	1,500*	4,076	0

* estimate ** "Royal Monaco" in 1977

Specifications

Engines: Polara

Types: aluminum block ohv slant six & cast iron ohv V-8
Six: 225ci (3.40 x 4.13), 145hp standard Monaco Six 1970–71
V-8s: 318ci (3.91 x 3.31), 230hp (150 net) optional 1967, standard 1968–72 & 1973 except wagons
360ci (4.00 x 3.58), 275hp (170–175 net) optional 1971–73, standard 1973 wagons
383ci (4.25 x 3.38), 270hp standard 1967; 290hp optional 1968–69; 275/300hp optional 1971; 325/330hp optional 1967–69
400ci (4.34 x 3.38), 190/250hp (net) optional 1972; 185/220hp (net) optional 1973
426ci (4.25 x 3.75), 335hp optional 1971
440ci (4.32 x 3.75), 335hp optional 1971; 350hp optional 1967–68, 1970 & 1969 wagons; 375hp optional 1968, optional except on wagons 1969; 235/285hp (net) optional 1972; 215/220hp (net) optional 1973

Engines: Monaco

V-8s: 318ci (3.91 x 3.31), 150hp (net) standard 1975–76; 145hp (net) optional 1977
360ci (4.00 x 3.58), 170–175hp (net) standard 1972, standard 1973 except wagons, optional 1976; 180hp (net) standard 1974, 200hp (net) optional 1974; 180/190hp (net) optional 1975; 155hp (net) standard 4dr 1977, optional 2dr 1977
383ci (4.25 x 3.38), 270–290hp standard 1967–71, optional 1967 500; 325–335hp optional 1967–71, standard 1967 500
400ci (4.34 x 3.38) 175–250hp (net) optional 1972–74, standard on wagons 1973; optional, standard on some models 1975–77
426ci (4.25 x 3.75), 335hp optional 1972
440ci (4.32 x 3.75), 350/375hp optional 1967–69; 350hp optional 1970; 230/280hp (net) optional 1972; 220hp (net) optional 1973; 230hp (net) optional except wagons 1974; 250hp (net) optional wagons 1974; 195/215hp (net) optional 1975; 205hp (net) optional 1976

Chassis and Drivetrain

Unibody construction
Transmission: In 1967–72, three-speed manual, TorqueFlite automatic optional, four-speed manual or TorqueFlite automatic on 1967–68 Monaco 500s; automatic standard from 1973
Suspension: torsion bar front suspension, solid rear axle with leaf springs

Measurements

Wheelbase (in): 1967–75: 122; 1976–77: 122.5; 1974–76 wagons 124
Curb Weight (lbs): 3,800–4,500
Tire size: 1967–68: 8.45 x 14, 8.25 x 14 (Polara 318); 1969: 8.25 x 14; 1968–69 wagons: 8.55 x 14; 1970–71: H78-15, J78-15 (wagons); 1972–73: F78-15 (Polara), G78-15 (Monaco), J78-15 (6 pass. wagons), L84-15 (9 pass. wagons); 1974–77 J78-15, LR78-15 (wagons)

Performance

(1967–71 383 V-8s)
Acceleration, 0–60mph: 12
Top speed (mph, approx.): 110
Fuel mileage (range): 12-16

Chapter 19

Fun	Investment	Anguish
	GTS	
9	7	3
	Other Models	
6	6	4

Collectible Darts

1967–76

History

The Dart compact was one of Dodge's all-time success stories. From the time it debuted in 1963 to the last one off the line in 1976, it was a consistent money spinner with a fine record of repeat business from customers who admired its practical size, clean styling, and the combination of performance and economy offered by its famous Slant Six.

But contemporary success stories do not breed collectible cars. Indeed, with certain exceptions (like the 1955–57 Chevy), collectibility seems usually to be enhanced in cars that sold in low numbers. The bulk of Dart production was six-cylinder sedans of no special pedigree: manifestly cars with no interest to collectors. What's important here is the low-volume sporty Dart—and, happily enough, there was at least one of these for every model year from 1967 to the end.

The early Dart shape that had grown so familiar (Chapter 15) vanished with the ambitious 1967 restyling program, which produced the best looking compacts yet. This was also an opportunity for Dodge to pare the Dart line, eliminating body styles that didn't sell (like station wagons and low-end convertibles) while catering to the popular taste for sports car attributes in a small, economical package.

The Dart GT continued at the top of the 1967 line, offering convertible and two-door hardtop bodies with the best grade of trim: vinyl bucket seats, padded instrument panel, full wheel covers, and special identification. As before, there was a choice between two Slant Sixes or the smallblock V-8. Adding the optional four-speed manual gearbox to a 225 Slant Six produced a nimble, sprightly performer; opting for the V-8 produced a nimble, impressive performer. Other options included front disc brakes, headrests, and a console-mounted tachometer (which was hard to see but better than none). Such

As popular as ever, the nimble Dart went through several good-looking styling periods in the late sixties and seventies. Only a handful of models have any collectibility, including this 1967 GT two-door hardtop. Look for this one with the 225 Slant Six and manual transmission.

A guide to what to look for: the 1968 Dart GTS (above), with bumblebee stripes, dummy hood scoops, divided grille, racy wheels, Red-streak tires and "GTS" badge, came with a handling suspension and the 340ci V-8; the more mannered GT-Sport (below) had either a 318 V-8 or the Slant Six. Convertible versions of both exist as well.

niceties only served to improve a fine basic package.

I've always tended to prefer small, fast cars to large, fast cars, and I confess a real liking for these Darts. I'm not alone. Testing a 235hp V-8 with the four-speed, *Motor Trend* recorded 0–60 in nine seconds flat and an impressive 16.5-second, 86.5mph standing quarter mile. Out of curiosity they drag-raced a Charger 383: "The Dart stayed about even in 0–30, 0–45 and 0–60, then whupped it handily in the quarter." Incidentally, here's a piece of nostalgia: the full price of *Motor Trend's* 235hp convertible was $2,860, about $9,000 in modern money. I don't think you can begin to buy that kind of performance for so little today. (In fact, that same Dart in top condition will cost you upwards of $12,000 today.)

During 1968–69, the Dart GT continued with little change, though the 235hp small-block option was replaced by a 318 V-8 developing about the same horsepower. A yet more esoteric confection in these years was the GTS or GT Sport: a hardtop or convertible packing Chrysler's hot new 340 V-8 and 275hp and much else besides: stiff suspension, E70-14 Red Streak Wide Oval tires, tail stripes, an engine dress-up kit with lots of brightmetal and crackle finish parts, and a choice of four-speed or automatic transmission. Forty more horses brought the acceleration times down to the eight second range and the firm suspension made the GTS a handler to boot. This was the nicest development of the breed and has my personal vote as the most desirable Dart ever built.

A handful of GTS Darts were sold with the big-block 383 V-8, and a few 1969 (hardtops only) were even equipped with the 440 engine. During the seller's market of the late eighties, the odd 383 or 440 would record a blockbusting auction sale, and though prices have settled somewhat, most price guides still rate them at a premium.

Since the GTS cost upwards of $3,500 as typically equipped, Dodge had the happy idea of dropping a 340 into a cut-price hardtop, creating the 1969 Swinger 340 at a base price of only $2,831. The Swinger was detrimmed compared to the GT, but it went like a squeezed lemon pit, and its stiff suspension (standard) gave it equal handling. In 1970, when the GT and GTS were dropped, it was the only performance Dart available.

The Swinger name went on a conventional line of Darts in 1971, but the high-performance engine was kept available on a new fastback coupe called the Demon 340, mounted on a new, shorter wheelbase. Demons featured prominent hood bulges and tie-down pins. They were continued in 1972 and again in 1973, when they were known simply as Sport 340s. Maintaining their level of performance despite emissions mandates, Dodge switched to a 360 V-8 for 1974–76, which was still putting out over 200hp in its final year.

Identification

All new, rounded bodies with a full-width grille and single headlamps were the rule for 1967: a handsome, chiseled look that was pleasing from any angle. Grilles were thin vertical blades with a center medallion in 1967, blacked out with a prominent horizontal bar and round parking lights in 1968; a thinner bar and thin oblong parking lights in 1969. GTSs carry "power-bulge" hoods and tail stripes. Convertibles were dropped in 1970; Swinger 340s had double hood scoops and a divided grille with vertical bars. Similar in 1971, the performance Dart now carried "Demon 340" identification and prominent fore-to-aft bodyside racing stripes with a matte black, twin-scoop hood. The boy-racer hood was eliminated in 1972 and the grille again adopted a horizontal theme. For 1973, the grille was cross-hatched and came to a central point. For the first time that year, a $435 sunroof option was offered; the '74 was identical except for a new taillight treatment, rear bumper and rear valance. A thinner textured grille with slim inboard parking lights identifies the '75s. Aside from a bodyside trim molding the '76 was virtually identical.

The1969 Swinger 340 was a detrimmed GT with no less the performance. The 1969 Darts were little changed, except for grillework. In 1970, when the GT and GTS were dropped, the Swinger 340 was the only high-performance Dart in the line.

Appraisal

The GTS is preferable to its 1970s successors on two counts: quality of fit and finish was much better in '60s models, and government regulations began to intrude on performance in the later V-8s. Furthermore, the GTS offered a convertible, Swingers and Demons only as coupes. Find a GTS with the four-speed and go out and tackle the nearest decreasing-radius turn: you'll see what I mean. These are fine road cars, able to coexist with what now passes for gasoline, packing bags of entertainment, and they still don't cost a king's ransom. However, if performance is your priority and you don't mind a coupe body, '70s models cost considerably less.

I've driven big-block GTSs and I don't find them fun. Performance is terrific off the line, but the cars are nose heavy understeerers which don't stop particularly well. From an investment standpoint they've done well, however.

The raw power of the 340 engine was comparatively unaffected by 1970s detuning, undertaken to reduce emissions in line with federal mandates. The 1973 version, for example, was still producing 240 net hp, and its 1974 replacement added enough cubic inches to maintain that power level. Even in 1976, the Dart's final year, the 360 was putting out 220 net hp. So this is one performance engine that can be relied upon no matter what the model year.

Summary and Prospects

Figures lie and liars figure, but look at how V-8 Darts have appreciated over the years! The hardtops have come up faster than the convertibles. In fact, if you bought a prime GTS hardtop in 1982 and sold it for top dollar in 1988, you realized a 38 percent return on your investment. Beats Krugerands! Most of the gains were made during the late 1980s, of course, when muscle cars of all kinds set repeated records. GTS values have stagnated over the last six or seven years. That doesn't mean they're stuck in place forever, and if the economy doesn't fall on its face you can predict that they will be in the $20,000-plus range by the end of the decade. Right now, they cost a bundle less than larger counterparts like Coronet R/Ts. The best buys, therefore, are the 1960s models, and 1970s models as recommended alternatives.

Six-cylinder GTs sell for rather less, but

The Swinger name went on a line of conventional Darts for 1971; the performance model was now the Demon 340, tricked out with hood scoops, rallye stripes, wide tires and blacked-out hood. Dodge built 10,098 Demon 340s this year and the supply remains pretty good.

if you find one with the larger 225 Slant Six, give it some consideration, especially if it has the four-speed manual gearbox. I'd take that combination over anything with an automatic. There's nothing wrong with TorqueFlite, but this is a driver's car.

A 1974 Dart Sport ("Demon" was dropped after 1972). By now the hot option was a 360ci V-8, which was still developing more than 200hp (net). Note, however, that from 1972, Dart Sports come with all sorts of engines, sixes included.

Price History

95+ point condition 1	1982	1988	1995	Return
1967–67 GT hardtop*	$1,800	9,000	12,000	17.1%
1967–69 GT convertible*	2,500	10,000	14,000	15.4%
1968–69 GTS hardtop	2,000	14,000	15,000	18.2%
1968–69 GTS convertible	2,750	18,000	18,000	16.9%
1968–69 GTS 383/440	3,000	20,000	20,000	17.1%
1969–73 340	2,100	6,500	7,500	11.1%
1974–75 360	2,000	5,000	6,000	9.5%

* Deduct one-third for six-cylinders

Production

	1967	1968	1969	1970	1971
GT	38,225	26,280	20,914	0	0
GTS	0	8,745	6,702	0	0
Swinger/Demon 340	0	0	n.a.	13,785	10,098

	1972	1973	1974	1975	1976
Demon/Sport 340	8,750	11,315	0	0	0
360	0	0	3,951	1,043	1,000*

* estimate

Specifications

Engines

Types: aluminum block ohv slant six & cast iron ohv V-8

Sixes: 170ci (3.40 x 3.13), 115hp standard 1967–69 GT six 225ci (3.40 x 4.13), 145hp optional 1967–69 GT six

V-8s: 273ci (3.63 x 3.31), 180hp standard 1967–69 GT V-8; 235hp optional 1967 GT V-8

318ci (3.91 x 3.31), 230hp optional 1968–69 GT

340ci (4.04 x 3.31), 275hp (240 net) standard 1968–69 GTS, 1970 Swinger 340, 1971–72 Demon 340, 1973 Sport 340

360ci (4.00 x 3.58), 245hp (net) standard 1974 360; 230hp (net) standard 1975 360; 220hp (net) standard 1976 360

383ci (4.25 x 3.38), 300hp, optional 1968–69 GTS

440ci (4.25 x 3.75), 340hp, optional 1969 GTS hardtop

Chassis and Drivetrain

Unibody construction

Transmission: three-speed manual, TorqueFlite automatic or four-speed optional; choice of TorqueFlite or four-speed on GTS

Suspension: torsion bar front suspension, solid rear axle with leaf springs

Measurements

Wheelbase (in): GT, Swinger 340 111; GTS 110; Demon 340, Sport 340 and 360 108

Curb Weight (lbs): GT hardtop 2,750; GT convertible 3,100; Swinger and Demon 2,900; 1973–76 340/360 3,300

Tire size: GT sixes 6.50 x 13; GT V-8s 7.00 x 13; GTS, Swinger, Demon and Sport 360 E70-14

Performance

(V-8s: 1967 GT 235hp / 1969 GTS 275hp)

Acceleration, 0–60mph: 9 / 8

Top speed (mph, approx.): 115 / 120

Fuel mileage (range): 13-19 / 11-17

Fun	Investment	Anguish
8	8	5
	Hemi	
10	9	10

Charger
1968–75

History

The fad for fastbacks in the 1960s lasted about as long as it had in the 1940s: five or six years. To its credit, Dodge was one of the first companies to recognize that the fad had ended, giving the axe to its Charger fastback only two years into production. The new 1968 Charger managed to be one of the most beautiful cars of the decade, while retaining the fastback's high-performance tradition.

William Brownlie, who had styled the original fastback, started with a clean sheet of paper for the new '68, drafting a svelte, slippery profile with a clean front end, smoothly curved body sides and a flowing roofline (one of the few that really looks good with a vinyl skin). The elegant rear deck with its faint spoiler stood out as unique and inimitable. "The only 1968 car which comes close to challenging the new Charger for styling accolades is the new Corvette," said *Car and Driver*: ironic, since

That little badge on the door says "Hemi," and designates one of the most desirable Chargers ever built. The all-new '68s were indeed one of the finest automotive designs of the sixties, smooth and clean, aggressive looking yet practical, with acres of glass and not a bad angle anywhere. The sky was the limit on Charger Hemis awhile back, but prices have lately come down to the realm of the sensible.

Dodge built close to 100,000 Chargers in 1968, most of which were powered by V-8s. A clean one with a modest 318 or 383 shouldn't set you back more than $15,000, and they can often be found for well under that. Unless you're shopping for Hemis or big-blocks, the earlier the better; emission standards made annual inroads on performance after 1970, and to the eyes of many the earlier cars are the best looking.

the '68 Corvette aged rapidly, while the Charger still looks good today.

While Chargers could be had with the Slant Six in 1968, most were V-8 powered. The R/T was the hottest, with a 375hp 440 Magnum as standard, or the Hemi optional. The 440 delivered standing quarter miles in the 14 second, 100mph range, actually better than some testers reported for the Hemi, but performance varied with final drive ratios.

The 1969–70 Chargers (including about 500 per year equipped with the Slant Six) were little changed in design or equipment. Also available was an SE (Special Edition) option comprising leather-faced bucket seats, simulated walnut steering wheel, pedal dress-up kit, light group, deep-dish wheel covers and map/utility pockets in the door panels.

An extreme facelift for 1971 resulted in a car that actually seemed bulkier than its real measurements, which were already ample. Though the wheelbase and rear overhang were trimmed, width and weight increased. The '71 looked like a combination of Pontiac front end and midsize Ford side elevation, with a humpy look over the rear wheels. The nimble elegance of the 1968–70 Charger had disappeared. Defying wisdom, however, the '71 sold better: over 82,000 units compared to about 50,000 of the previous vintage.

Four different trim stages now gave the Charger six different permutations. After the baseline coupe (fixed rear side windows and B-pillar) and hardtop (both with Slant Six standard) came the V-8s: the 500, SE (with handsome customized roofline) Super Bee, and R/T. The Super Bee, like the Coronet with that designation, had rock-hard suspension, heavy-duty brakes, three-speed floorshift, needle instruments, oversize battery, and a 275hp 383 V-8 that burned regular gas. Optional colors included such vivid varieties as Citron Yella, Green Go, Plum Crazy and Hemi Orange. Strategically priced, the 'Bee started under $3,300, about $500 below the R/T, "a great looking piece of man's iron," as Dodge put it, "that knows how to live on a budget." Both the Super Bee and R/T could be ordered with the 440 Magnum and Hemi, the 440 being standard R/T issue. Even with the Hemi, a fully outfitted 'Bee cost under $5,000, a performance bargain. The problem

The 1971 Charger R/T sold for about $4,500 as typically equipped. R/Ts featured a body-colored front bumper, exposed headlamps, blacked-out grille, tape stripes, and simulated door louvers.

was that by the early 1970s, insuring one of these fire-breathing monsters could cost that much for only two years—the insurance companies were out to kill the muscle car.

Just as rapidly as Bryan Nichols had changed the Charger from a fastback to a muscle car, Bob McCurry changed it into a personal luxury car. By 1972 the Super Bee, R/T, and Hemi vanished, leaving only the coupe, hardtop, and SE, and advertisements for "personal luxury" Chargers.

As a result, the Charger remained a success. Production didn't drop out of sight like that of so many super-car rivals; model year 1973, with the same basic drivetrains and a clumsy facelift, set the all-time production record of nearly 120,000 cars. Even during the oil boycott year of 1974 Dodge managed to flog nearly 75,000 Chargers, which was hardly bad considering that it looked pretty much the same as '73.

Identification

Restyled for 1968 as a semi-fastback hardtop coupe with recessed grille, coke bottle fuselage and round taillights in black valance panel. The '69 used a divided grille while the '70 grille carried a horizontal center bar. R/Ts can be told by their badges but not always by "bumblebee" stripes running across the rear deck; these were optional on all Chargers, as were longitudinal stripes on R/Ts in 1969–70. Restyled again in 1971, the Charger had heavier, rounded lines, a divided bumper/grille and a heavy "hump" over the rear wheel wells. Headlamps were no longer hidden and the grille bars horizontal in 1972. The 1973 Charger had a crosshatch grille pattern while the '74 reverted to horizontal bars with a heavy center bar. SEs had a stand-up hood ornament and opera windows to go with the landau vinyl roof.

Appraisal

The performance examples cited in our specifications speak volumes about the dwindling of Charger muscle in the 1970s. But everything is relative, and as Jim Brokaw wrote in 1972, even the later Charger was "the best of super-car survivors. The butt-grabber buckets are placed to give you a near perfect driving position, response is as good as can be had, and with that combination of front torsion bars and rear leaf springs of the correct dimension, augmented by anti-roll bars, one can envision a number 11 or 71 on the door. . . ."

Collectors place great store by styling, however, and here the 1968–70 generation is clearly superior to the 1971–74. The 1971 changes increased shoulder room and trunk space slightly, but as A.B. Shuman wrote,

R/T and Super Bee were replaced in 1972 by the Charger Rallye, a package including special needle instruments, heavy-duty suspension, and 340ci four-barrel V-8. "Slotted" doors identify the Rallye.

"those positive aspects somehow don't seem to offset the feeling that you're wheeling around in one of those full-sized behemoths rather than a sprightly intermediate."

On later Chargers, look for the optional Rallye package (standard on all 440s): heavy-duty suspension with jumbo sway bars front and rear, F70-14 whitewall tires, needle instruments, louvered taillights, power bulge hood, and fuselage extractor vents.

Summary and Prospects

Despite what we've said about the aesthetic merits of the 1968–70 Chargers, the '71s have clearly done better in value gains over the past ten or twelve years. Although none of them have gained in value since 1988, and the Hemis have actually dropped, over the long haul they've still returned very handsomely on investments. The consensus seems to be that performance is more important than styling: the heavy 1971 facelift hasn't affected the esteem of collectors as much as the severe downgrading of sporting models and performance engines starting in 1972.

Through 1971, Chargers are desirable cars that should remain solid investments for the foreseeable future.

The 1973 Charger SE (Special Edition), the most popular '73, at a time when the Charger's image was changing from performance to personal-luxury. Vinyl top, SE badges identify these models.

Price History

95+ point condition 1	1982	1988	1995	Return
1968–70 Charger V-8	$4,800	17,000	17,000	11.1%
1968–70 R/T	5,250	20,000	20,000	11.7%
1968–70 R/T Hemi	7,000	40,000	27,500	12.1%
1971 Super Bee	2,600	18,000	18,000	17.4%
1971 R/T	2,900	20,000	20,000	17.4%
1971 R/T Hemi	4,000	40,000	27,500	17.4%
1971 SE	3,000	19,000	19,000	16.6%
1972–74 coupe	2,300	6,000	6,000	8.3%
1972–74 SE	3,100	7,000	7,000	7.0%

Production

	1968	1969	1970	1971
Charger	96,108	69,142	39,431	46,183
R/T	*	20,057	10,337	3,118
500	0	**	*	11,948
Super Bee	0	0	0	5,054
SE	0	*	*	15,811

* production combined with Charger ** see next chapter

	1972	1973	1974
coupe	7,803	11,995	8,876
hardtop	45,361	45,415	29,101
SE hardtop	22,430	61,908	36,399

Specifications

Engines

Types: aluminum block ohv slant six & cast iron ohv V-8
Six: 225ci (3.40 x 4.13), 145hp (105–110 net) optional 1968–74
V-8s: 318ci (3.91 x 3.31), 230hp (150 net) standard 1968–74
340ci (4.04 x 3.31), 240hp (net) optional 1972–73
360ci (4.00 x 3.58), 200/245hp (net) optional 1974
383ci (4.25 x 3.38), 290/330hp optional 1968–70, 275hp standard 1971 Super Bee; 275/300 optional 1971
400ci (4.34 x 3.38), 190/255hp (net) optional 1972; 175/260hp
(net) optional 1973; 185/205/250hp (net) optional 1974
426ci (4.25 x 3.75), 425hp optional 1968–71 R/T
440ci (4.32 x 3.75), 375hp standard 1968–71 R/T, optional 1968–71; 390hp optional 1970 R/T; 385hp optional 1971 R/T; 280hp (net) optional 1972–73; 330hp (net) optional 1972; 275hp (net) optional 1974

Chassis and Drivetrain

Unibody construction
Transmission: three-speed manual, TorqueFlite automatic or four-speed manual optional
Suspension: torsion bar front suspension, solid rear axle with leaf springs

Measurements

Wheelbase (in): 1968–70: 117; 1971–74 115
Curb Weight (lbs): 3,100–3,600
Tire size: 1968–70 sixes: F78-14; 1968–70 V-8s: 7.35 x 14; 1968–71 R/T and 1972 Rallye: F70-14 (some F70-15); 1971–74
E78-14

Performance

Acceleration 0–60mph (secs.): R/T Hemi 7.6; R/T 440 7.2; 1971
Super Bee Hemi 5.7; 1972 440 Rallye 8.2; 1973 440 SE 7.5
Standing-start quarter-mile:
1968 R/T Hemi: 15.35 seconds @ 94.6mph
1970 R/T 440: 14.71 seconds @ 96.7mph
1970 Charger 440 Six Pack: 13.95 seconds @ 101mph
1971 Super Bee Hemi: 13.73 seconds @ 104mph
1972 440 Rallye: 16.2 seconds @ 89mph
1973 440 SE: 15.7 seconds @ 88mph
Top speed (mph, approx.): R/T Hemi 130; R/T 440 115; 1973 440 SE 120
Fuel mileage (range): R/T Hemi 9-15; R/T 440 11-17; 1973 440 SE 10-12

Chapter 21

Fun	Investment	Anguish
	Charger 500	
9	9	9
	Charger Daytona	
10	10	8

Charger 500 & Daytona

1969

History

On stock car ovals the new-generation Charger was more slippery than the 1966–67 Charger fastback, but its recessed grille and inset backlight slowed it down. A Hemi-Charger qualified at 185mph for the 1968 Daytona 500, but Cale Yarborough's fastback Mercury Cyclone did 189, and at those speeds 4mph meant the length of a football field every lap.

So the Dodge Boys came up with a special '69 Charger, suitably modified for racing and appropriately called the 500. Designed for improved airflow, it had a flush grille and backlight, a higher decklid spoiler, and a 2in lowering job. Dodge then built just over 500 to qualify as "stock" according to National Association of Stock Car Auto Racing (NASCAR) rules.

Hot Rod dazzled readers with a quarter-mile run of 13.48 seconds at 109mph in their Daytona 500 test car, which seemed to augur well, but at the far higher speeds of NASCAR competition the 500 was a bust. Aero-engineer Gary Romberg told this writer, "We didn't do enough to the shape.

Charger Daytonas and Plymouth Superbirds at a winged warrior convention at the Chelsea Proving Grounds.

A flush grille and backlight, designed for improved airflow, quickly distinguishes the 1969 Charger 500. Dodge built just over 500 to qualify this model as "stock" for NASCAR, but competition performance was underwhelming and back they went to the drawing boards.

We just tipped it at the right rake angle and used a more slippery backlight." At the 1969 Daytona 500, Lee Roy Yarbrough's Ford won going away. Dodge needed to do more. Their solution, released during the summer, was the memorable Charger Daytona.

Engineer John Pointer described the Charger Daytona's wind-cheating features to sales vice president (and future general manager) Bob McCurry: "a shark-like nose, a Charger 500 back glass and biplane wings mounted on two oversized 1957 Plymouth fins." Pointer also made a hasty sketch. McCurry hardly gave it a glance: "If it'll win races, build it!"

Dodge duly built 505 Charger Daytonas to qualify them as "production" with NASCAR. Though dealers took 1,200 orders, the build run was strictly to qualify, and dealers were told to install disappointed customers in something else. To those who were fortunate enough to obtain one, the 1969 Charger Daytona cost $4,000. Allegedly Chrysler lost $1,000 to $1,500 on each one, but winning races, not making money, was their purpose.

The racing Daytona ran a Hemi V-8 and a close-ratio four-speed with Hurst shifter. Driver Bobby Isaac said it idled "like a coffee can full of rocks [but] as far as acceleration is concerned, the Hemi sure turns on where the others shut off." One late modification added what looked like air extractors to the front fenders. Actually, they were blisters added to prevent tire scrub during hard cornering. Daytonas were built by Creative Industries, the famous Detroit builder of specials and prototypes including the famous Packard Panther Daytona (a limited edition whose name had nothing to do with any intent to race).

The Charger Daytona's first competition appearance was at Talladega in September 1969, with Richard Brickhouse setting a new official world's closed-lap speed record of 199.996mph during qualifying, and breaking 200 in the race. Much to Dodge's disappointment, Ford chose to sit out that race, but a month later at Charlotte, they walked all over the Chargers, which had to ease off to preserve their tires. According to Rathgeb, "Firestone could not—and Goodyear would not—build a tire that could stand up to 200mph. After five laps you were out of rubber."

Finally at the Texas 500 in December, Bobby Isaac's Daytona toppled the Fords with a 144.277mph average. Over the course of the 1969 season, Charger Daytonas won 22 Grand Nationals, four fewer than Ford. Unfortunately that proved both the first and last year for the winged Charger. Late in 1969, Plymouth wooed Richard Petty back from Ford, and the aero men set to work building a winged 1970 Plymouth, the Superbird, for the following NASCAR season. Nevertheless, thirteen drivers on five major circuits compiled a record of eight-two Dodge victories in 1969.

Identification

The 1969 Charger 500 should not be confused with 1970–71 500s, which were simply up-market versions of the base Charger. The racing-specific '69 500 can easily be identified by its flush backlight, flat grille, low stance, and racing tail strips surrounding big "500" numerals on the rear fenders. The Charger Daytona used this same body with a pointed nose, hidden headlamps, and huge tailfins supporting a horizontal wing for high speed stability. Street Daytonas (and many racing ones) had their fins, wing, and tail painted contrasting colors with the "Daytona" name dropped out in body color in big capital letters.

Appraisal

Rockets off the line and stable at virtually any speed, both Super-Stockers are tremendous road cars and enormous fun to drive. The penalties are a thirst for high-octane fuel and certain body parts and glass unique to these models only, always a hazard for restorers. Such singular cars will always have relatively limited appeal, but for those who can appreciate their great character, the Super-Stockers will always be among the greatest cars Dodge ever built.

Summary and Prospects

Our 1988 prices below are only an average, because there were very wide deviations. Contemporary price guides had Day-

Full-race Charger Daytona ran Hemi 426s mated to close-ratio four-speeds with Hurst shifters. Their first appearance, at Talledega in September 1969, saw Daytonas circling the track at a hair under 200 mph during qualifying.

This is a Daytona production prototype at the Proving Grounds. The dummied-up front end and that nose delineation is actually nothing more than black plastic tape.

The aggressive, low and pointed nose and enormous spoiler worked to keep the beast hunkered down at NASCAR speeds. Racing Daytonas copped twenty-two Grand Nationals in 1969, but this was four fewer than Ford, who had started earlier.

The prototype posed for photographs. Production totaled 505, and a remarkable number are still around. Prices for show-worthy examples are all over the map, the high limit at present being $50,000, but do shop around.

tonas down around $22,000, while auction bidders sometimes approached six figures for Hemis. *Automotive Investor*, the collector's financial newsletter, doesn't list the 1969 Charger 500 but rates the Daytona as a "hold," anticipating negative return on investment for the next several years and giving a price range of $35,000 to $39,000 for "high 2 condition." The value of these very special Chargers is almost wholly determined by supply and demand; since only about 500 of each were built, it seems logical to assume that demand will always permanently exceed supply by a large margin. Therefore, they remain very good investments. On looks alone, the Daytona is destined always to sell for substantially more than the 500.

Price History

95+ point condition 1	1982	1988	1995	Return
1969 Charger 500	$4,800	25,000	30,000	16.4%
1969 Charger Daytona	6,500	40,000	50,000	18.5%

Production

Charger Daytona production was exactly 505
Charger 500 production was at least 500 but the exact figure is not stated by Dodge.

Specifications

Engines

Types: ohv V-8
426ci (4.25 x 3.75), 425+hp optional
440ci (4.32 x 3.75), 375hp standard

Chassis and Drivetrain

Unibody construction.
Transmission: four-speed manual, Hurst linkage
Suspension: torsion bar front suspension, solid rear axle with leaf springs

Measurements

Wheelbase (in): 117
Curb Weight (lbs): 3,700
Tire size: Officially F70-14 but larger tires were routinely fitted

Performance

(Street models)
Acceleration, 1/4 mile (four-speed): 13.5 seconds @ 109mph
Top speed (mph, approx.): 135
Fuel mileage (range): 9–13

Chapter 22

Fun	Investment	Anguish
7	7	5
	Hemi	
9	9	10

Challenger
1970–74

History

For a long time Chrysler Corporation had a me-too reputation: it would wait until rivals had established new markets and then lunge in, usually too late. But this was not always true of Dodge Division, which was often the only sign of life in the stagnating corporation during the 1960s. Dodge moved relatively quickly with market adjustments: the upsized Dart compact (1963), the Charger fastback (1966), and the all-new Dart line (1967). Dodge was definitely slow to wake up to the "ponycar," the long hood, short deck, big engine sporty concept popularized by the 1965 Ford Mustang. Lowly American Motors had a Mustang-competitor two years before Dodge.

One reason for Dodge's delay was the dichotomy between the Dodge and Plymouth compacts, from which their ponycars stemmed. While the 1965 Valiant converted easily into the Barracuda, the Dart didn't,

Low, fast, and aggressive looking, the 1970 Challenger R/T convertible is one of the most desirable seventies Dodges. Unfortunately its market timing was off, and it did not survive long. Convertible production was extremely limited; only 1,070 like these were built.

After over 12,000 R/T hardtops for 1970, sales for the '71 model were disappointing at 4,630, and the R/T was dropped for 1973. All R/Ts were hardtops, and all were V-8s. Hemis broke the $50,000 mark in the late 1980s, but have lately settled down around $30,000 for prime examples.

and Dodge dealers didn't see a market for Valiant-based "glassbacks" with Dodge badges.

Finally, when Plymouth finally produced a ground-up Barracuda ponycar for 1970, Dodge seized the opportunity to produce its own version. Although physically related, however, the only common parts to the two models were the window glass, and Challenger was built on a longer, 110in wheelbase. Dodge product namers had shunned fish and fowl for the nice, gutsy "Charger" label on their mid-sixties fastback; here again they avoided Mother Nature monikers and picked the name "Challenger" for their new ponycar.

It was an appropriate label, because challenge is exactly what Dodge faced. Challenger was intended as a larger, more luxurious Mustang with a price to match. This immediately limited its appeal to a fraction of the ponycar market—which, by 1972, was slowing fast. Dodge had jumped in too late. By 1975 only the Mustang and Camaro ponycars had survived, while Challenger, Barracuda, and AMC's Javelin were out of production. Still, the Challenger outsold the Barracuda in every year of production.

Challenger opened with an ambitious, broad line in 1970: standard and R/T convertible, coupe, and SE. The latter ("Special Edition") was a luxury coupe with vinyl top, small backlight and deluxe interior. R/Ts

The only styling change for the 1973 Challenger was the 5mph bumper, but the line had been greatly pared back: convertibles, R/Ts, and SEs hadn't been around since 1971, and the Rallye, though still available, was only a trim package. Still, Dodge managed to build over 32,000 '73s.

A 1973 Challenger Rally can be distinguished by its blacked-out grille. Expect to pay about $12,000 for the very finest examples; most sell for well under five figures.

were the performance jobs, with a 383 V-8 standard, packing heavy-duty suspension and brakes, raised letter blackwall tires, "rallye" instruments, longitudinal or bumblebee tape stripes, and individual badging. Curiously, you could order an "R/T SE," which seems like a contradiction in terms, and is not common. Challengers were also available with the Hemi (the Dodge equivalent of the Hemi 'Cuda), and the jumbo 440 V-8 was available on R/Ts.

Expanding the line even further in early 1970, Dodge added two special models: the Challenger "Deputy" starting at only $2,700, and the Challenger T/A, designed for Sports Car Club of America (SCCA) Trans-Am sedan racing (where it finished a distant fourth). The Deputy, with fixed rear side windows and the small 198ci Slant Six as standard, was a belated response to the Challenger's original high price, but the T/A was far more interesting. Standard T/A equipment included the 340 Six Pack V-8, sport suspension, and four-speed manual gearbox.

Dodge began pulling back on Challenger offerings in 1971, when the convertible was restricted to the base Challenger line (it was dropped altogether in 1972, along with the R/T). The '72 lineup was restricted to base Challenger and Rallye coupes, the latter distinguished by their heavy- duty suspension, front disc brakes, and stiff stabilizer bars front and rear. Horsepower was being expressed in net rather than gross. Rallyes came with the 240hp 340 V-8. Equipped as standard with the three-speed manual gearbox, Rallyes could be ordered with the four-speed or TorqueFlite.

Identification

Compared to the similar-appearing Barracuda, the Challenger has quad instead of single headlamps, a prominent body side crease and a longer wheelbase. A wall-to-wall grille cavity incorporates the quad headlamps, with a single, oval, central grille outline in 1970, divided into two sections for 1971. For 1972 the grille was a black oblong within the front cavity (now painted body color); for 1973 large black bumper guards were added to meet federal crash standards. There were no changes to the 1974 model, which must be identified by serial numbers.

R/Ts had chrome bumpers in 1970. In 1971, bumpers were color-keyed, and R/Ts also have simulated brake cooling scoops in front of the rear wheels and new tape treatments. No longer was an "R/T SE" available, and the convertible was confined to the base Challenger, which did not offer the big engines, probably in response to insurance concerns for overpowered ragtops. For 1972–74 the sportiest model was the Rallye, distinguished by its "performance" hood, dual exhausts, simulated front fender air extractors, and special badging.

Appraisal

On styling and performance grounds the Challenger has a lot to recommend it, but potential buyers should consider the downside, and whether it matters. *Car and Driver* summed this up in a 1970 road test: "There's no doubt the Challenger is a handsome car but it also has a massive feeling—which stems from a full 5in more width than a Mustang and a need to sign up with Weight Watchers—that is totally unwelcome in a sporty car." While Challengers are perfectly acceptable for routine transportation, they won't get down a bit of wiggly road with finesse. They understeer and tend to be overpowered. The Hemi has never been fun in traffic (*C&D's* example had bad throttle response, backfired through the carburetor, idled rough and sometimes quit altogether.)

Like the Barracuda, Challenger parts are not easy to find, especially body and interior pieces; engine and drivetrain parts are much easier, because they are so often interchangeable with Coronets and Chargers.

Summary and Prospects

As with the Coronet R/T and Charger, conventional wisdom has it that big-engined models (1970–71 in the Challenger's case) are the cars to get. This fails to take into account the commensurately lower prices of 340, 383, and even Slant Six models (1970–72), which have returned about the same compound annual rates as the big-blocks. From the standpoint of prestige and collectibility, Six Pack and Hemi Challengers are obviously the most desirable; but you may find a 383 or 440 four-barrel a lot more affordable. The point is that past value appreciation of these Dodges suggests that a Challenger is a good buy *whatever* type it is. Watch out in all cases for modifications, especially mechanical, which greatly lower the value of these cars.

Big-engined cars tend to have received lots of abuse over the years, and purchasers should be aware of and check for potential trouble. Paint color can significantly affect value: Plum Crazy or Hemi Orange are big plus factors, dark green metallic a negative; however, black, which is ordinarily not much in demand, is so far that it can raise the value by as much as $5,000, according to *Automotive Investor*: "But if the build sheet is missing, beware of bogus fender tags. If both are missing, it is next to impossible to prove the original color or document all factory options."

The last of the line, the 1974 Challenger Rallye, distinguished by its door and fender trim and (again) by the blacked-out grille. A small fraction of the 16,000 '74 Challenges had the Rallye package, and their usual power was a 318 V-8, though some can be found with the 340.

Price History

95+ point condition 1	1982	1988	1995	Return
1970–71 base convertible	$2,600	8,500	14,000	15.0%
1970 R/T convertible	3,000	9,500	18,000	16.0%
1970 T/A coupe	3,000	9,500	17,500	15.8%
1970–71 R/T coupe	2,500	8,000	13,000	14.7%
1970–71 R/T Hemi coupe*	6,000	55,000	35,000	15.8%
1972–74 Rallye coupe	3,000	6,500	12,000	12.2%
1972–74 base Challenger	2,300	5,000	10,000	13.0%

*Hemi prices are highly erratic. Some owners of Hemi convertibles claim they're worth $175,000; they are worth, in fact, whatever someone is willing to pay, but it's not $175,000. Not now, anyway.

Production (approx.)

	1970	1971	1972	1973	1974
base hardtop	53,337	23,088	18,535	32,596	16,437
Rallye hardtop	0	0	8,123	*	*
convertible	3,173	2,165	0	0	0
SE hardtop	6,584	0	0	0	0
R/T htp (incl T/A)	12,489	4,630	0	0	0
R/T T/A hardtop	2,400	0	0	0	0
R/T convertible	1,070	0	0	0	0
R/T SE hardtop	3,979	0	0	0	0

* Rallye production included with base hardtop 1973–74

R/T production	1970	1971
440 Six Pack convertibles	99	0
440 Six Pack hardtops	1,936	250
Hemi convertibles	9	0
Hemi hardtops	347	71

Specifications

Engines

Types: aluminum block ohv slant six & cast iron ohv V-8

Sixes: 198ci (3.40 x 3.64), 125hp standard, Deputy coupe 1970–71

225ci (3.40 x 4.13), 145hp standard all other sixes 1970–72

V-8s: 318ci (3.91 x 3.31), 230hp (150 net) standard 1970–74

340ci (4.04 x 3.31), 265hp (240–245 net) optional 1970–74;

275hp optional 1971

383ci (4.25 x 3.75), 275–300hp optional 1970–71 R/T

440ci (4.32x 3.75), 370–390hp, optional 1970–71 R/T

Chassis and Drivetrain

Unibody construction

Transmission: three-speed manual, four-speed or Torqueflite optional

Suspension: torsion bar front suspension, solid rear axle with leaf springs

Measurements

Wheelbase (in): 110

Curb Weight (lbs): 3,150–3,500

Tire size: 7.35 x 14, F70-14 on R/T or Rallye

Performance

(340/425 V-8s with 3.55:1 axle ratio)

0–60 secs. 1/4-mile secs. @ mph

R/T Hemi 5.814.1 @ 103.2

440 4bbl. 6.414.8 @ 95.0

440 Six Pack 6.015.7 @ 91.2

Chapter 23

Fun	Investment	Anguish
3	2	8

Charger to Mirada
1975–83

History

Bowing to the virtual end of the muscle car era, Dodge retargeted the Charger as a personal-luxury coupe and introduced it in 1975 as a clone of the Chrysler Cordoba, from which it was hard to tell at a distance. Like the favored coupe of Ricardo Montalban, Charger SE featured a long hood, short deck, and smaller round parking lights mounted Jaguar-like, well inboard from the headlamps, flanking a square grille. Base engine was the old reliable 318; V-8s up to 400ci were optional. The lookalike 1976 model included a Daytona trim package, first use of the name since the winged marauder of 1969: but in this case it stood simply for a $345 option comprising two-tone paint, special body tape, dual sport mirrors, and color-keyed bodyside moldings. A Brougham package of luxury goodies was also available and fitted

It may have been a sheep in wolf's clothing, but the 1978 Magnum sure looked good. The clean, Cord-like grille and glass-covered headlamps gave it an aerodynamic appearance ahead of its time. The public seemed to agree, since 90 percent of Dodge's personal-luxury sales in 1978 were Magnums.

The '76 Charger Daytona, a far cry from the 1969 version, was a a $345 trim package comprising two-tone paint, special body tape, dual sport mirrors and color-keyed bodyside moldings. About 7,500 were built.

to nearly half of the SEs. The "Charger" name also went onto a number of standard hardtops, noncars of little interest to collectors; the one we're concerned with is strictly the SE.

A power sliding sunroof and T-bar roof with removable tinted-glass panels were optional for the first time on the 1978 version.

A much swoopier looking and, in its way, rather elegant variation was the 1978 Magnum SE, actually a replacement for the Charger SE which was then in its final year. Magnums had an "aero" front end with glass-covered headlamps (illegal in most states; they swiveled out of the way then the lights were switched on) and a simple grille reminiscent of the classic Cord 810. Magnum ran another year, then gave way in 1980 to a new name, Mirada.

This 1978 Charger SE is an extremely rare car: only 2,735 were built, and they represent the very last of the Charger line. The bulk of sales by now were Magnums. From a rarity and historical standpoint this is a Dodge worth looking for, but spending more than $5,000 is a mistake.

The Mirada wasn't just a nameplate change: commensurate with the new age of lightness and federal requirements for corporate fleet economy, it was 400lb lighter and 6in shorter than the Magnum and its base engine was the Slant Six, but an optional touring package called the "S" type offered the 360. Mirada production dropped off rapidly after 1980, and Dodge produced the last of the breed in 1983.

Identification

Thin bar at top of grille for 1975; thick for 1976. Grille for 1977 was a black crosshatched pattern divided into three horizontal sections; for 1978, the Charger SE's last year, the crosshatches had chrome highlights. Magnums are immediately identifiable by their glass-covered headlamps and Cord-like slotted grilles; the 1979 models had two-segment wraparound taillamps. A really awful two-tone pattern put contrasting colors in sections on front and rear fenders; with the right colors it almost resembled a tank camouflage paint scheme. Magnums carry that name in block letters on front fenders.

Mirada grilles were composed of four wide slots separated by five bright horizontal strips which bent back at the base in 1980. Prominent spoke wheels and hydraulic lifters for the Slant Six were featured in 1981. For 1982 the hood carried twin creases which came forward to frame the grille sides. Appearance was unchanged in 1983 except for new colors.

Appraisal

Look for Magnum's Grand Touring package: color-keyed fender flares blending into body, heavy-duty shocks, GR60-15 tires, leather-wrapped steering wheel, firm-feel power steering, and special colors. Mirada's more compact dimensions and lighter weight make it a better driver's car, and to my eyes it's also better looking. The most important variation is the first-year (1980) model with the optional 360ci engine—the only Mirada with real go-power. Also look for Miradas with the CMX package, the Sport Equipment package, and goodies like aluminum road wheels. Do not buy anything that looks to need replacement body or interior parts because these are very scarce. Avoid rusters, too.

Summary and Prospects

These cars were still depreciating ten years ago, and their return on investment figures are skewed by this factor. They've now risen to the $5,000–6,000 level, which is where you can expect them to remain for a long time; do not anticipate 10 percent return on your money if you buy one today. Their appeal is visceral and subjective. Magnums and Miradas to my eyes look terrific for anything that big, and much more up-to-date than the corresponding Chrysler Cordoba. But whether they are truly collectible is a debatable point. The ones to buy, as in any other recent production of this nature, are those with exceptionally low mileage in mint condition.

A swoopy-looking 1982 Mirada CMX, one of only 1,474 built for that model year, with convertible-style top, turbo sheels, and slotted fenders. This is a sharp piece of styling that still looks good today, but whether it will ever become serioiusly collectible is debatable.

Price History

95+ point condition 1	1985	1988	1995	Return
Charger SE	$1,500	4,200	4,500	11.6%
Charger SE Daytona	1,800	4,500	5,000	10.8%
1978 Magnum	2,000	4,700	5,300	10.2%
1981 Mirada (V-8)	4,500	2,000	6,000	2.9%
1983 Mirada (V-8)	6,600	3,000	6,200	-0.6%

Production (approx.)

	1975	1976	1977	1978	1979
Charger SE incl. Brougham	30,812	34,872	30,979	2,735	0
Charger SE Daytona	0	7,295	5,225	0	0
Magnum	0	0	0	46,966	23,697
Magnum GT	0	0	0	861	1,670

	1980	1981	1982	1983
Mirada	27,1651	1,899	6,818	5,597
Mirada "S"	1,468	0	0	0
(CMX package on above)	(5,384)	(1,683)	(1,474)	(1,841)

Mirada packages: Only 981 1980 Miradas had the convertible type roof (only seven were the "S" type); for 1981 the convertible type roof was part of the CMX luxury-touring package. Production of the CMX package is listed above.

Specifications

Engines

(net hp)

Types: aluminum block ohv slant six & cast iron ohv V-8

Six: 225ci (3.40 x 4.13), 90–110hp standard 1980–83

V-8s: 318ci (3.91 x 3.31), 135–150hp standard 1975–79; 120hp optional 1980–83

360ci (4.00 x 3.58), 170–190hp optional 1975–79; 130/185hp optional 1980

400ci (4.38 x 3.38), 175–185hp optional 1975–78

Chassis and Drivetrain

Unibody construction

Transmission: three- or four-speed manual standard, TorqueFlite and five-speed manual optional

Suspension: torsion bar front suspension, solid rear axle with leaf springs

Measurements

Wheelbase (in): Charger SE/Magnum 115; Mirada 112.7

Curb Weight (lbs): Charger/Magnum 3,950; Mirada 3,250 (avg.)

Tire size: GR78-15 (1975–77); FR78-15 (1978–79); P195 75R-15 (1980–83)

Performance

(318 V-8)

Acceleration, 0–60mph: 11–12

Top speed (mph, approx.): 100–105

Fuel mileage (range): 12–18

Chapter 24

Collectible Omnis

1979–86

Fun	Investment	Anguish
	024, Charger, Rampage	
5	3	6
	Shelby Charger	
7	6	6
	Omni GLH	
8	2	6

History

A watershed year for Dodge, 1978 saw the last of the really big land yachts and the first of Dodge's new state-of-the-art compact, Omni, cloned with Plymouth's Horizon and designed at last to do battle with the hordes of Japanese small cars which had been eating Detroit's lunch for half a decade. With a 99in wheelbase, the shortest in Dodge history, Omni packed a new transverse-mounted overhead cam 1.7 liter four (built by Volkswagen—there's an irony) and standard four-speed manual transaxle, with optional TorqueFlite. It was destined to last a dozen years, and in many of those it made the difference between profit and loss.

In 1979 came the derivative Omni 024, coupe, a neat little bundle of energy when teamed with the 2.2 liter engine (optional from 1981). Earlier 024s were no great performance shakes, but one of possible collector interest is the de Tomaso option, which included a sport suspension, P185/70R13 belted tires on cast aluminum road wheels, vinyl bucket seats, and leather steering wheel.

This book is not about Dodge trucks, but a pickup that falls in the gap between car and truck is worth mentioning: the pretty little Rampage, built mainly to use up excess capacity at the Belvidere, Illinois plant which

I remember these "Omni People" from outer space (the Bronx, probably) at the 1978 New York Automobile Show. They wandered around asking folks to go have a look at the new Dodge Omni—which was considerably more down to earth. But Omni provided a platform that would spawn several interesting and possibly collectible Dodges in the years ahead.

How much influence Signore deTomaso had over the 1980 Omini DeTomaso is a good question, but the package included a sport suspension, P185/70R13 belted tires on cast aluminum road wheels, vinyl bucket seats, and leather steering wheel, which made it a cut above the ordinary.

Charger 2.2 revived a famous name on the L-body coupe: a tape-stripe package with extra instruments and a rorty exhaust (1983) which was fun to drive when mated to the five-speed gearbox. (The four-speed ratios are too far apart for real gear-stirring fun.)

assembled 024 coupes. Unique in being the only pickup besides VW's Rabbit with front-wheel drive, it failed to sell. Only about 27,000 were built over three years: an incredibly low figure for Detroit. As a result, Rampages are uncommon today. They look great (admittedly, this writer has owned two), hold half a ton, and are fun to drive with the four- or five-speed gearboxes.

In 1983 the O24 designation for Omni coupes gave way to a name with some tradition: Charger, with Charger 2.2 the up-market item. This could be teamed with a new five-speed overdrive manual gearbox as alternative to the standard four-speed. Then in midyear came the sporty Shelby Charger, named for the famous racing driver and builder of Cobras and Shelby-Mustangs, now in collusion with his friend Lee Iacocca at Chrysler. Standard on the Shelby was a front air dam, rear roof appliques, special paint and trim; the high-output 2.2 engine was standard equipment, as was a *very* tight sports suspension and bigger tires and wheels.

The Omni itself offered a collectible variation in 1984–86 called the GLH: a semi-limited edition with Q-ship performance: *Collectible Automobile* compared it to "the 98lb nerd in the old Charles Atlas ads who muscled-up to take revenge on that sand-kicking beach bully." The GLH was fathered by Carroll Shelby, who suggested a Shelby-ized Omni to Iacocca. Shelby also named it: GLH for "Goes Like Hell."

Omni GLH (for Goes Like Hell) was designed by Carroll Shelby to fool superior types in far more expensive iron. For $7,500 you got blacked-out trim, 129lb-ft of torque, close-ratio five-speed manual transmission, quick ratio power steering, power front disc brakes, "Swiss-cheese" alloy wheels, and Eagle GT tires on 15in wheels. Well driven, a GLH would do 0-60 in 8.5 seconds and corner around 0.8g, which was pretty impressive.

Dodge built about 25,000 Rampage pickups between 1982 and 1984. A car-based truck like the El Camino, it didn't sell, but its rarity, lithe proportions, and spirited front-wheel-drive performance rate it a minor collectible. Good ones cost as little as $2,500. (The '84 has a much cleaner front end than the busy '83 shown here.)

The 1984 Omni GLH carried the Shelby Charger's 110hp engine, which gave 129lb-ft of torque (compared to only 87 in ordinary 1.6 liter Omnis). A five-speed manual transmission with close-ratios, quick-ratio power steering, power front disc brakes, "Swiss-cheese" alloy wheels and Eagle GT tires on jumbo 15in wheels completed the spec. GLHs did 0–60 in under nine seconds and cornered around 0.8g, a figure equivalent to the Corvette; yet they started well under $7,500, well under that of the rival Volkswagen GTI.

Better news was coming: the 1985 GLH sported Dodge's new Turbocharged 146hp engine from the Daytona Z sports coupe, packing 168lb-ft of torque and 0–60 times down around seven seconds, yet returning 18 to 20 miles to the gallon, and starting at only $7,620. Shelby added still more power in 1986 with the GLH-S ("Goes Like Hell-Somemore), produced in his own shop in California: the ultimate Omni, mayhap the ultimate compact Dodge. Similar performance was, of course, available on the Charger coupes, but they *looked* the part. The Omni GLH looked perfectly respectable until you put your foot down.

Charger (and Shelby Charger) production ended in March 1987 when the Belvidere, Illinois, plant was closed to change over to the new Chrysler New Yorker and Dodge Dynasty—good cars but eminently noncollectible.

Identification

All models can be identified by their badging. Coupes and Rampages through 1983 use a grille composed of six slots flanked by single headlamps, blacked out on de Tomaso models. Quad rectangular headlamps and a twin-oblong air intake arrived on 1984 Chargers and Rampages. Omni GLHs carry GLH badges and the special equipment mentioned above.

The 1984 Shelby Charger, conjured up by Carroll Shelby was a more genuine specialty sports model than the earlier DeTomaso. Specs included front air dam, rear roof appliques, special paint and trim, high-output 2.2, and very stiff, Shelbyesque suspension.

Appraisal

Like Shelby's Mustangs, his Chargers and Omni GLH were crude but effective. They rode like farm carts, hard loud exhausts, and were notchy to shift. Their most notable characteristic was that without torque steer they were a handful to drive but they went precisely like hell. It's great fun to embarrass performance iron with the Omni GLH: to this day, nobody ever sees it coming! Rampage pickups are nose heavy and may need some weight in the bed to hold them down on rough roads. Omnis coupes and Rampages are rust resistant, but when they go they go fast; curiously, one of the unlikely places for tinworm is the A-pillars, where you don't really need any corrosion.

Assembly quality varies considerably on these cars: some came out of the factory as tight as a drum, others very sloppy. The driving position is ungainly, with the seatbacks reclined too far, even in the full-upright position (inserting a washer or two under the back seat bolts can alleviate this). Look for the five-speed, which has much better-spaced gear ratios and is easier to shift than the four-speed and more fun than the automatic.

Summary and Prospects

The compound rates of return calculated below cannot be relied upon to reflect long-term appreciation, since all of these cars were still depreciating in 1989 and their values were taken from used car data books, not collector car price guides. Therefore the figures you see are purely for comparison. It's worthy of note, however, that prime examples of all models (and remember these are 95-point show-winners, not your typical used cars) have already turned around and begun to appreciate in value. They have a long way to go to make their original prices ($7,000–9,000, more if you factor for inflation), but the fact that they *have* turned around so soon seems significant. It will be interesting to look at their values five years from now, when, hopefully, it will be time for an updated edition of this book! For now: consider them reasonably good, low-buck investments. Also, do not pay this kind of money for anything except very well-preserved originals, and avoid any premium-priced example that is claimed to be "restored."

Shelby Turbo Charger from 1985 had more suds, thanks to 146hp turbo engine borrowed from the Daytona. Special Shelby paint job, Swiss cheese alloy wheels, and blanked out rear quarters distinguished this fine-performing Omni derivation.

Price History

95+ point condition 1	1989	1995	Return
1982 Omni 024	$2,000	4,500	14.4%
1980–81 de Tomaso	1,800	4,000	14.2%
1982–84 Rampage	1,500	3,500	15.1%
1984 Charger 2.2	3,000	4,600	7.4%
1983 Shelby Charger	2,725	5,500	12.5%
1985 GLH Turbo	3,800	4,200	1.7%

Production

	1979	1980	981
Omni 024	57,384	61,650	41,056

	1982	1983	1984	1985	1986
Charger	0	22,535	34,763	38,203	38,172
Charger 2.2	14,420	10,448	11,949	10,64	54,814
Shelby Charger	0	8,251	7,552	7,709	7,669
Omni GLH	0	0	3,285	6,513	3,629
Rampage	17,067	*7,500	*3,000	0	0

*estimate

Specifications

Engines

(net hp)

Type: overhead cam four

1.7 liters, 105ci (3.13 x 3.40, 63hp standard 1979–86

2.2 liters, 135ci (3.44 x 3.62), 84–85hp optional 1981–82; 94–96hp optional 1983–85; 110hp standard Charger 2.2 and Omni GLH; 146hp (turbo) standard 1985–86 Shelby, optional 1985–86 Omni GLH

Chassis and Drivetrain

Unibody construction

Transmission: four-speed manual standard; TorqueFlite automatic optional; five-speed manual optional starting in 1984

Suspension: MacPherson struts with coil springs and anti-sway bar front; trailing-arm independent suspension with coil springs and sway bars rear

Measurements

Wheelbase (in): 99.2 (Omni), 97.6 (coupes), 104.2 (Rampage)

Curb Weight (lbs): 2,200 (coupes), 2,300 (Rampage)

Tire size: P175/75R13, P155/80R13

Performance

Acceleration, 0–60mph: 8.7 (110hp), 7.0 (Turbo), 12 (85hp)

Top speed (mph, approx.): 115 (Turbo)

Fuel mileage (range): 17–25

Chapter 25

Fun	Investment	Anguish
6	2	8

Convertibles

1982–86

History

The initial success of its survival-effort K-car convinced Chrysler Corporation to produce a variety of follow-up models using the same platform. Indeed for several years, critics would charge that all Chrysler offered was variations on the K. While this was essentially true, the variations were well-planned and timed to perfection—the basis of Chrysler's recovery in the 1980s. One such example was the convertible, which Lee Iacocca set out to revive after noting public reaction to a prototype he drove around Detroit. The proto, like the production ragtops which followed, were built out of K-body coupes by a specialty firm, Cars and Concepts, Inc.

The Dodge 400 convertible debuted along with its Chrysler LeBaron counterpart in 1982. It was the first Dodge softtop since 1971. The convertibles were making money by 1984 (when the model name became

Convertibles were happily revived under Lee Iacocca's management (though despite his claims he had little to do with it personally). This is the second-year, a 1983 400. Ragtops were K-car based with local Detroit conversion by Cars and Concepts Inc.

Dodge's ragtop reached peak popularity in its final year of production, 1986, when over 12,000 600s (like this shot) and over 4,000 600ES models were built. These cars are still depreciating, so the thing to do is keep an eye on used car lots and grab that creampuff when it comes along at a giveaway price. Best time to buy is just before winter!

"600") and had amortized their production costs by the time this generation of intermediates was replaced by the Lancer in 1987. Later, Dodge would offer convertible body styles in the new Shadow series (see Chapter 30).

The most desirable of 1980s convertibles is the 1984–86 600ES Turbo, powered by the then-most-potent version of Chrysler's 2.2 liter "Trans-4" engine. Like other ES models, it had the "sport/handling" package: thick front and rear sway bars, "high-control" shocks, "firm-feel" quick-ratio power steering, and beefy tires on multihole cast alloy wheels. Aside from its wheels, the ES Turbo looked much the same as conventional 600 ragtops, but with less brightmetal and a standard black-finish luggage rack mounted on its deck.

Except for 1984 when a manual five-speed gearbox was optional, convertibles were equipped with TorqueFlite automatic transmission. Interiors were standard K-car, in the 400/600 LeBaron style, save for a three-spoke sport steering wheel on the ES. All models had the same unfortunate, hard-to-read digi-graphic instrumentation, and, for 1983–84, annoying "voice" warnings for everything from low oil pressure to "door ajar." The power top had a glass backlight and separate rear-quarter windows.

Identification

The 400 and 600 are badged in the style of Mercedes-Benz and had a slat-type grille similar to that of the Mirada, along with two vertical bodyside louvers in the front fenders. Aside from individual badges, the ES has a black finish deck rack and multihole alloy wheels; although these were technically optional on other models, they were standard on the ES.

Appraisal

The major driveability problem on these cars is cowl and body shake. The top works fine and Turbos are quick, yet reasonable on gas; handling is good and the size about right for four or maybe five passengers. Excess throttle lag and a lot of heat under the hood are typical problems for turbos. Most owners will disconnect the stentorian-toned vocal warning system, a real piece of Mickey Mouse which lasted about as long as drivers took to listen to its repertoire. Parts and service are not a problem, and quality of fit and finish is very good.

Summary and Prospects

There is no sign that these cars are going anywhere as investments, but the ES Turbo will definitely be collectible in years to come, especially if the convertible body style again goes out of favor. Turbos are the best bet. Sinking big bucks into one of these is not recommended, but don't buy a basketcase, either: they are complicated automobiles, and restorations are not cost-effective.

Price History

All 1982–86 convertibles are still depreciating, so there's no point in listing a price history. Current market values range between $2,500 and $3,500 for prime examples, with the ES Turbo toward the top of the range and the early 400 model (1982–83) toward the bottom. An exceptionally clean 600ES with low mileage might bring $6,000, and is worth it.

Production

	1982	1983	1984	1985	1986
400/600	5,541	4,888	*10,960	8,188	11,678
600ES	0	0	*	5,621	4,759

*estimated ES production in 1984 about 2,000 units

Specifications

Engines

(net hp)

Type: overhead cam four

2.2 liters, 135ci (3.44 x 3.62), 84–85hp standard 1982; 94–
96hp standard 1983–84; 97–99hp standard 1985–86;
146hp (turbo) optional 1984–86
2.5 liters, 153ci (3.44 x 4.09), 100hp optional 1986
2.6 liters, 156ci (3.59 x 3.86), 92–93hp optional 1982–85

Chassis and Drivetrain

Unibody construction

Transmission: four-speed manual standard; TorqueFlite automatic optional; five-speed manual optional starting in 1984

Suspension: MacPherson struts with coil springs and anti-sway bar front; beam axle, trailing-arms, coil springs and sway bars rear

Measurements

Wheelbase (in): 100.3
Curb Weight (lbs): 2,500–2,600
Tire size: P185/70R14

Performance

Acceleration, 0–60mph: 8.7 (110hp), 7.0 (Turbo), 12 (85hp)
Top speed (mph, approx.): 115 (Turbo)

Chapter 26

Fun	Investment	Anguish
6	3	8
	Turbo/Shelby/IROC	
8	4	8

Daytona

1984–93

History

One of Chrysler's most interesting derivations on the K-car platform was the G-24 sports coupe, which had a long lead time during which prototypes amazed the automotive press, so interesting was its "international" specifications. The Daytona, and a Chrysler version called Laser, had a much modified K-chassis featuring all-coil springs, front MacPherson struts, trailing arm rear beam axle, and power rack-and-pinion steering topped by a husky 2+2 "fasthatch" body with design overtones of the Porsche 928. Introduced twelve years ago, they still look good today. The Daytona's styling was further improved with a major facelift in 1987: disappearing headlamps and an ultra-smooth fascia. The Laser was dropped that year so the Daytona had a monopoly with the corporation and sales soared. A mild facelift in 1989 provided a smoother front end and wraparound taillights.

The hottest Daytonas were the 1984–86 Turbo Z, 1987–88 Shelby Z, 1988–91 Daytona

In its early years the new Daytona looked like an other-worldly car and few believed it came from Chrysler Corp., which had been rumored to be at death's door just a few years before. Turbos with Swiss cheese alloy wheels are the most desirable models.

Daytona interiors are businesslike ergonomic design exercises (this is an '84), with such luxury touches as firmness adjustment (note rubber squeegie at driver's seat) and dashboard-mounted systems monitor. Lever action reclining seats were a retrograde feature, however.

Shelby and 1992–93 IROC R/T. The "Z" designation apparently traded on the image of the Camaro Z28: Carroll Shelby was, of course, Chrysler's performance consultant; "IROC" stood for "International Race of Champions," where the Daytona competed. All were powered by a transverse turbocharged 2.2, initially the 146hp "Turbo I" with five-speed or optional automatic. Horsepower increased on the 1987 "Turbo II" (manual shift only) via an intercooler and higher boost pressure; internals were suitably strengthened to cope.

By 1990 the "Turbo IV" 2.5 liter engine and a new interior were available, along with a V-6 for the IROC model. The super-performer IROC R/T stayed with the turbocharged four, which had now evolved to

The 1986 Daytona Turbo Z, like its 1984-85 predecessors, was powered by a transverse turbocharged 2.2, the 146hp "Turbo I," teamed with a five-speed manual or optional automatic transmission. The only problem with buying these is that they tend to have led hard lives, so check the potential purchase carefully, especially if offered at a giveaway price.

the sophisticated 16-valve 2.2, which had previewed a year earlier in the Spirit R/T (see Chapter 29).

Called "TwinCam Turbo," this potent four produced 224hp, just over 100hp per liter—the highest hp/displacement ratio of any production engine in Chrysler history. Its one-piece cylinder head, pistons, and intake/exhaust manifolds were developed with the aid of Lotus (which was nice of General Motors, Lotus's owner). Mated to this torquey powerplant was a close-ratio five-speed manual transmission with high-capacity gearset "for better precision and durability." It was the only gearbox available on the IROC R/T.

On plainer Daytonas, a Mitsubishi-made 3.0 liter V-6 was made available in 1990, along with a high-performance turbocharged 2.5 liter four. Front suspension was recalibrated to improve steering feel and straight-line stability, and a compact disc player option brought back memories of Chrysler's old-time "Highway Hi-Fi" (but with far superior sound).

The Dodge Daytona is one of those happy designs that lives a long time in public esteem. It lasted through 1993, a decade of production.

Identification

The original base Daytona lacked a spoiler, but '85 models had a body color three-piece rear spoiler similar to that of the '84 Turbo Z. The '87s were facelifted with a hidden-headlamp "droop snoot" and revamped tail. A driver's airbag became standard in 1988, wraparound taillights and all-disc brakes in 1989, a new dash and cockpit-adjustable variable-rate shocks for 1990. On base models a C/S Competition Package made available most items provided as standard on Shelbys, including a hot engine, beefed-up chassis, racy exterior add-ons, and lower weight. The '89 Shelbys lost their previous "Z" suffix and had unique paint treatment graduating from charcoal in the lower panels to body color in the upper panels. For 1990 front end was restyled with twin horizontal slots and had new wraparound taillights.

A fabulous performer in the Scat-Pack tradition, the 1988 Daytona Shelby Z represented one of the high points for the long-running Daytona sporty car series. Moonroof, slatted backlight, airdam, deep bucket seats, and distinctive trim made for an exciting package when mated to the fine turbo engines.

Appraisal

A tight, agile suspension and tremendous performance from Zs, Shelbys, and IROCs is balanced by a stiff ride (especially on Shelbys), lots of noise, and clunky manual shifters. Quality of fit and finish was below par through the 1986 models, but better than average from 1987 on. If you want a decent ride, less noise, fewer boy-racer trim items but nearly the same performance, consider the Pacific model (1987–88) in lieu of a Shelby. Thanks to its long life the Daytona has benefited from many improvements, so the later the model, the better the overall driving package.

Summary and Prospects

It is too soon to make long-term predictions about the collectibility of these Daytonas: interesting specifications in the Zs and Shelbys are balanced by the coupe body style, and since so many are produced there is no noticeable scarcity. It does appear that the IROC models and R/Ts especially will definitely have collector value ten or fifteen years from now. Scarcely 1,000 were built over two years, and no more than 300 were

R/Ts. From an investment standpoint a Daytona is worthwhile if it is a mint original with low-mileage. Likewise, a degree of collectibility is almost certain for anything labeled Shelby, and the Turbo is an exciting performer for the car collector who (in contrast to the typical used car buyer) is willing to look around for that rare mint original and pay more than the book says it's worth. There may also be some residual "last car" interest in the '93 models, which had the smallest production run of any model year.

Price History

Daytonas are still depreciating, so a price history is superfluous. Current used car market values for pre-1990 models range between $1,500 and 3,000, but collectors pay up to $5,000 for extremely fine Z-cars. The used car market determines values of more recent Daytonas and is changing (depreciating) them rapidly at present; consult latest editions of N.A.D.A. *Official Used Car Guide.*

Non-Shelbyized Daytonas are also worth considering, especially limited editions like this 1990 "Spring Special," with special paint, trim, and wheels. It'll be some time yet, if ever, before Daytonas start heading back up the price scale, and there are plenty to choose from, so shop around.

Production*

	1984	1985	1986	1987	1988
Daytona	44,347	31,473	29,166	30,120	67,272
Turbo/Shelby Z	5,000	8,023	17,595	7,152	7,580

	1989	1990	1991	1992	1993
4cyl	0	21,569+	10,773	5,607	4,369+
4cyl turbos	80,878	3,614+	1,329	593	181+
V-6	0	13,265+	8,341	7,278	4,477+

*1984 breakdowns are estimates. Also, breakdowns for 1990 and 1993 are sales and therefore somewhat below actual production. Total production for 1990: 43,785. Total production for 1993: 9,677.

1992 model production: IROC total 310 (estimate 100 R/Ts)
1993 model production: base 5,167; ES 2,952; IROC 725; IROC R/T 181

Specifications

Engines

(net hp)
Type: overhead cam four and ohc V-6
Fours: 2.2 liters, 135ci (3.44 x 3.62), 97–99hp standard 1984–86; 146hp (turbo) optional 1984–88, standard 1987–88 Pacifica; 174hp standard 1987–88 Shelby; 224hp (16 valve) IROC R/T 1992–93
2.5 liters, 153ci (3.44 x 4.09), 96–100hp optional 1986; standard 1987–89; 150–52hp (turbo) optional 1989–91, standard Daytona Shelby 1989–91
2.6 liters, 156ci (3.59 x 3.86), 150hp optional 1990
V-6:3.0 liters, 181ci (3.66 x 2.87), 141hp optional 1990–93; standard IROC 1992–93

Chassis and Drivetrain

Unibody construction
Transmission: four-speed or five-speed manual standard; TorqueFlite automatic optional; Turbos with automatic never received more than the 146hp engine.
Suspension: MacPherson struts with coil springs and anti-sway bar front; trailing-arms with coil springs and sway bars rear

Measurements

Wheelbase (in): 97.1
Curb Weight (lbs): 2,600–3,000
Tire sizes 1984–89: P185/70R14; P205/60R15 (1985–86 Turbo, 1988 C/S package); 195/70R14 (1987– Turbo); 225/50VR15 (1988 Shelby, 1989 C/S); 205/55VR16 (1989 Shelby)

Performance

Acceleration, 0–60mph (secs.): 8-9 (Turbo I); 6.5 (IROC R/T
Top speed (mph, approx.): 120 (Turbo I); 140 (IROC R/T)
Fuel mileage (range): 17–25

Chapter 27

Fun	Investment	Anguish
7	8	7

Shelby Dodges
1986–89

History

Shelby Automobiles Inc. in Whittier, California, was born again in early 1986 to write more performance history, this time with Dodge instead of Ford. Chrysler had formed the company as a subsidiary "skunkworks" to devise low-volume, high-performance offshoots of production Dodges, much as Shelby had built its version of the Ford Mustang during 1967–70. The partnership was based on the same people as twenty years before: racing driver-turned-builder Carroll Shelby, and Ford executive-turned-Chrysler-savior Lee Iacocca. Though entirely Dodge-based, these cars were badged and registered as Shelbys, albeit available through Dodge dealerships.

1986–87 GLH-S: The first new Shelby was a more specialized rendition of the 1983–86 Omni GLH (Goes Like Hell; see Chapter 24), itself a hotshot derivative of Dodge's workaday subcompact. Its main distinction was a more potent, 175hp version of the normal GLH's optional "Turbo 1" engine.

GLH-S can be quickly spotted by its decal trim and wild wheels. Shelby's touch improved the aerodynamics and tweaked its already potent engine up to 175hp, the makings of a real pocket rocket. About 500 were built.

Charger GLH-S used the same tuned engine and suspension as the five-door but the Charger coupe body for better aerodynamics, and as you might expect outsold the five-door two to one. Shelby claimed 0-60 in under 7 seconds, a 94mph quarter mile in 14.9 seconds, and a top speed of 135mph.

A modified plenum chamber, air-to-air intercooler, equal-length intake runners, and other Shelby tweaks accounted for the extra power. Also standard were a manual five-speed transaxle, heavy-duty suspension, wide tires on multihole alloy wheels, tachometer, console, and unique cabin trim.

1987 Charger GLH-S: This followed the lines of the GLH-S, being a Shelby version of Dodge's hatchback, related to the GLH-S in the same way ordinary Chargers relate to Omnis. It should not be confused with the 1982–87 Shelby Chargers (Chapter 24), low volume models built and sold as Dodges. Mechanically identical to the GLH-S five-door and similar in chassis specs and equipment, the Charger GLH-S was sportier looking and more aerodynamic.

1987–88 Shelby Lancer: A more luxurious model based on the H-body Lancer-five-door, this Shelby came with power everything, offered a remote-control CD audio system and leather interior as options, and featured Shelby's signature on a dashboard plaque, a special sport steering wheel and lacy-spoke road wheels. Its exterior "aero" package added a front air dam, side and rear skirts, and a rear spoiler to the already smooth Lancer shape. Suspension was stiffened per Shelby practice, which meant a very hard ride but superb front-drive handling, provided you kept a rein on the 175hp Turbo IV engine. Carroll liked to portray this as an American answer to the Mercedes 190E 2.3–16, which it wasn't. It cost much less, and it was far less sophisticated. Production was deliberately very limited; performance and not refinement was its reason for being, and Shelby sold every one he built.

1988–89 CSX: The basis for this Shelby was the new (for 1987) P-body Shadow, the Omni's replacement, which brought the series of Dodge-based Shelbys full circle. Two

Driving Lights, big "Lancer" decals on front doors and special wheels differentiate the 1987 Shelby Lancer from Dodge's Lancer Shelby (see next chapter). Of the 800 built, half came with automatic and half with a close-ratio five-speed manual gearbox.

Not just another Shadow: the Shelby CSX was the final and most sophisticated Shelby-Dodge, with a special suspension package, Shelby mag wheels and the tweaked 175hp powerplant. In production for 1988-89, it managed over 2,200 sales, and there are still quite a few around.

differences from the Omni: the CSX was a three-door not five-door hatchback, and much smoother and more modern looking. Because the production Shadow was derived from Chrysler's K-car compact like the Lancer, the CSX followed the same basic technical formula as the Shelby Lancer. But it was slightly faster owing to its smaller size and lower curb weight.

Identification

GLH-S Omni and Charger and Shelby Lancer are plainly different from the Dodge production versions, with large Shelby decals on body sides and tops of windshields. Shelby badges are used inside and out. The Shelby Lancer is different from the Lancer Shelby! (See next chapter.) The CSX is obviously different from stock Shadows through its aero-styled front end and is also Shelby badged and decaled.

Appraisal

Gloriously unrefined, almost crude in engine and chassis behavior (*Consumer Guide* said of the GLH-S, "the accelerator pedal acts like a lane change switch") the Shelby Dodges are fast, noisy, and fun. CSXs built after 1988 benefited from Chrysler's VNT—Variable Nozzle Technology—turbocharging system that all but eliminated throttle lag for more responsive low- and midrange acceleration.

Summary and Prospects

High performance plus low production and the Shelby mystique adds up to definite collector prospects for these cars in the not-too-distant future. The CSX has the distinction of being the most numerous Dodge-based Shelby special.

Price History

Shelby Dodges are still depreciating and will do so indefinitely, but the annual losses have been only about half the rate of normal Dodges. Consult current editions of used car value guides. Original list prices were $17,395 (GLH-S), $12,995 (Charger GLH-S), $17,000–18,000 (Lancer), and $13,495–15,000 (CSX)

Production

	GLH-S	Charger	Lancer	CSX
Approximate:	500	1,000	800	2,250

Specifications

Engines

(net hp)
Type: overhead cam turbocharged four
2.2 liters, 135ci (3.44 x 3.62), 146hp (GLH-S automatic) 174hp (five-speed manual)

Chassis and Drivetrain

Unibody construction
Transmission: five-speed manual; automatic optional on GLH-S
Suspension: MacPherson struts with coil springs and anti-sway bar front; trailing-arm independent suspension with coil springs and sway bars rear

Measurements

Wheelbase (in): 103.1 (GLH-S), 96.5 (Charger GLH-S), 103.1 (Lancer), 97.0 (CSX)
Curb Weight (lbs): 2,850 (GLH-S), 2,483 (Charger GLH-S), 2,895 (Lancer), 2,700–2,800 (CSX)
Length (in): 180.4 (GLH-S, Lancer), 174.8 (Charger GLH-S), 171.7 (CSX)

Performance

0–60mph (secs.): GLH-S 8.5, Lancer 7.7, CSX 7.1
1/4 mile (secs. @ mph): Lancer 15.7 @ 89, CSX 15.1 @ 90
Top speech (mph): 130–135

Chapter 28

Fun	Investment	Anguish
2	1	7

Lancer Shelby
1988–89

History

This was a sporty low-volume offshoot of Dodge's late 1980s front-wheel-drive Lancer sedan, the H-body five-door hatchback also sold as the Chrysler LeBaron GTS. Not to be confused with the Shelby Lancer (see previous chapter), an all-out version built by Shelby in California, the Lancer Shelby was based on the production Lancer ES models. To this Dodge added a monochrome exterior (first in red or white, then black), lower-body "aero" skirts, low-profile rear roof spoiler and a leather interior, plus an uprated "handling" chassis and intercooled 174hp "Turbo II" engine. A five-speed manual was the only transmission with this engine, though automatics were teamed with the 146hp "Turbo I." Air conditioning and a power driver's seat were standard.

Hold your thumb over the grille! Doesn't this 1988 Lancer Shelby look like a Saab 9000 CD? Well, the difference will probably be more obvious from behind the wheel. I doubt late model four-door sedans will ever be collectible, but one of these will provide fun transportation for a very modest price. (Lack of "Lancer" decals on front doors and rear roof spoiler distinguish it from the Shelby Lancer.)

Identification

Distinctive alloy wheels, blacked out grille reminiscent of the old Chrysler 300 Letter Series, distinctive rear roof spoiler, and special badging.

Appraisal

A roomy and versatile hatchback, the Lancer Shelby provides good performance with economy and is dirt cheap, if you can find one, on today's used car market. Against its positive qualities come mediocre quality of fit and finish, a clunky manual shifter, turbo boom, and lag.

Summary and Prospects

Collectibility is far from certain, given the existence of the Shelby-built Lancer with higher power and even lower production. Still, the Lancer Shelby is an interesting recent factory flyer that's quite practical for day-to-day transport, and sure beats the rank-and-file compacts. It was dropped with the other Lancers after 1989.

Price History

Lancer Shelbys are still depreciating and will do so indefinitely; consult current edition of used car value guides. Original list price was $17,395.

Production

Model year production was only 279 for 1988 and an unknown but probably similar figure for 1989.

Specifications

Engines

(net hp)

Type: overhead cam turbocharged four

2.2 liters, 135ci (3.44 x 3.62), 146hp (automatic) 174hp (five-speed manual)

Chassis and Drivetrain

Unibody construction

Transmission: five-speed manual or automatic

Suspension: MacPherson struts with coil springs and anti-sway bar front; trailing-arm independent suspension with coil springs and sway bars rear

Measurements

Wheelbase (in): 103.1

Curb Weight (lbs): 2,850

Length (in): 180.4

Chapter 29

Fun	Investment	Anguish
3	2	6

Spirit Turbo

1989–92

History

Dodge's version of Chrysler's new line of front-drive, four-door notchback sedans was a twin to the Plymouth Acclaim. Dodge called it "midsize": bigger than the compact Aries, about the same size as the 600 which it replaced. True to its performance heritage, Dodge offered a turbocharged ES model Spirit right from the start, and even today, Spirit V-6s with close to 150hp are available brand new off dealer floors.

While Plymouth tailored the Acclaim for family duty, Dodge tilted Spirit toward the sporty. Aside from the turbo engine, Spirits also offered a handling suspension, which gave superior response at some expense of ride comfort.

The 1991 model year saw a familiar designation reappear in the Spirit R/T, using a new performance engine never before offered in a Chrysler product: a 16-valve, double overhead cam, turbocharged 2.2 liter

While Plymouth's A-body new-era compact was angled toward family hauling, Dodge called its version the Spirit and tilted toward performance. Spirits carrying "Turbo" badges are little Q-ships. This is the 1989 prototype, shown outside Chrysler's Highland Park headquarters.

Differences are subtle: the 1991 Spirit R/T (top), with dohc, sixteen-valve turbocharged 2.2 and 224hp, can be identified by its rubrail badge and rear spoiler; the 1991 Spirit ES (bottom) with V-6 is a more standard, but by no means underpowered, version.

with 224hp and 217lb-ft of torque. This powerplant was later utilized by the exciting 1992–93 Daytona IROC R/T. Anti-lock brakes were added to the options list (and are well worth looking for), replacing the standard disc/drum combination with four-wheel discs.

Identification

Upright and boxy, these four-door sedans display little external differences from year to year; refer to serial numbers and registration documents.

Appraisal

The five-speed is somewhat notchy and demands a sure hand during critical gear changes. Spirits have front disc brakes and though they display some nosedive, control and pedal feel are excellent. It's a square and tallish little rig, but fun to drive with the high-performance engines. New car buyers spending upwards of $18,000 may have wished for a more flamboyant look than the Spirit R/T delivered; on the other hand, it proved to be a Q-ship in the same "spirit" as the old Hemi Coronets, and that's something.

Summary and Prospects

Real collectibility of the popular and plentiful 1989–90 Spirit Turbo is in doubt, but we mention this car as a dark horse, if only to point out Dodge's continuing concern with high performance halfway through the 1990s. The Spirit R/T is more interesting and a great little road car, and its numbers suggest collectibility. Turbo engines teamed with the five-speed (with all of its faults) can produce a great daily-driver at very little cost right now, and the 224hp R/T is a true performance Dodge.

Price History

Still depreciating and likely to do so indefinitely.

Production

	1989	1990	1991	1992
Turbo	16,432	3,816	0	0
Turbo R/T	0	0	1,172	787

Specifications

Engines

(net hp)

Type: turbocharged four

2.5 liters, 153ci (3.44 x 4.09), 150hp (ohc turbo, 1989–90)

2.2 liters, 135ci (3.44 x 3.62), 224hp (dohc turbo, 1991–92)

Chassis and Drivetrain

Unibody construction

Transmission: five-speed manual or automatic

Suspension: MacPherson struts with coil springs and anti-sway bar front; trailing-arm independent suspension with coil springs and sway bars rear

Measurements

Wheelbase (in): 103.3

Curb Weight (lbs): 3,060

Length (in): 181.2

Chapter 30

Fun	Investment	Anguish
6	5	5

Shadow Convertible

1991–92

History

The Dodge Shadow subcompact shared a basic platform with the Dodge Daytona, but offered a more upright design for better passenger and cargo space. Interiors were roomy and nicely packaged, with reclining bucket seats standard, though there wasn't much room in the back.

Dodge revived the convertible body style in the Shadow range five years after the last 600 ragtops had been run off. Topless, the Shadow attracted a lot of attention but not a lot of buyers. It was withdrawn after the 1992 model year. The convertible conversion was carried out by an aftermarket modifier, American Sunroof Company. It has a manual folding top with a plastic rear window and a two-place rear seat equipped with shoulder belts and, like all Shadows, the steering wheel contains an airbag.

There will probably come a time when the American industry will quit making convertibles again, because as much fun as they are, they sell no better now than they did thirty years ago. This 1991 Shadow is Dodge's most recent effort.

Spirit convertible for 1992, final year for the body style, can be identified by its multislot wheels with exposed lugs and special side decaling.

Identification

The only convertible in the 1991–92 Dodge line, part of the Shadow ES series.

Appraisal

Consumer Guide described the Shadow convertible was "a composed small car that doesn't have to be driven fast to be enjoyed. Cowl shake and body flex were evident, but not to a troubling degree."

Summary and Prospects

By nature all convertibles have a degree of collectibility. Production numbers for the Shadow were not high even during its initial sales spurt following introduction, and the number of '92s is intriguingly small. This is probably as good a bet for a future collector car as any conventional Dodge of its era, but paying top dollar for one is not warranted, given the years that will have to go by before it achieves artifact status.

Price History

Still depreciating; in 1995 retail prices were around $8,000 for nice examples, and falling fast. Original list in 1991 was $13,000 for the five-speed, $13,500 for the automatic, but most convertibles were optioned out to around $16,000.

Production

1991	1992
18,731	3,599

Specifications

Engines
(net hp)
Type: turbocharged four
2.5 liters, 153ci (3.44 x 4.09), 100hp standard; 150hp optional

Chassis and Drivetrain
Unibody construction
Transmission: five-speed manual or automatic
Suspension: MacPherson struts with coil springs and anti-sway bar front; trailing-arm independent suspension with coil springs and sway bars rear

Measurements
Wheelbase (in): 97.0
Curb Weight (lbs): 2,910
Length (in): 171.7

Chapter 31

Fun	Investment	Anguish
8	7	7

Stealth
1991–

History

Reaction to this new generation high-performance coupe was uniform, loud, and enthusiastic: "Think of it as an affordable Ferrari 348," wrote *Motor Trend* of the top-of-the-line Stealth R/T Turbo. A fabulous looker, Stealth traces its heritage to a 1988 concept car called Intrepid, now the name of Dodge's superb new "cab-forward" sedans, and an association with Mitsubishi, which builds a near copy called the 3000GT. Chrysler (in the person of Tom Gale and his crew of talented Highland Park designers) did the exterior styling, Mitsubishi the interior and mechanical hardware: the best of two worlds.

The Stealth platform is based on the Mitsubishi Eclipse/Plymouth Laser/Eagle Talon sports coupes built at the Chrysler-Mitsubishi plant in Illinois. Stealth and 3000GT share that trio's 97.2in wheelbase, but step up a class in size with a body that's 10in longer and 6in wider. The 1991 Stealth came in four models, all with standard driver's airbag and a transverse-mounted 3.0 liter V-6 engine. Base, ES, and R/T models were front-wheel drive. The base car used a 164hp V-6 with two valves per cylinder; the ES and R/T had a double overhead cam variant engine with four valves per cylinder and 222hp. The R/T Turbo added a pair of intercooled turbochargers for 300hp and a permanent four-wheel-drive system that can reallo-

It sure doesn't look like your father's Oldsmobile—or his Dodge, either. Sprung from the Intrepid showcar, the curvaceous Stealth was an immediate hit and a styling sensation. Not all of them went the way they looked but every one was a testimony to Dodge's continued quest for high performance and sports car excitement.

Stealth Twin Turbo for 1992; few changes occurred from year to year and the best clue is the wheels.

cate a normal 45/55 front/rear power mix as needed to maintain traction on slippery pavement. The R/T Turbo also has standard four-wheel steering that kicks in above 30mph and is designed to improve high-speed handling. Both R/Ts have body flares and scoops and a nonfunctional rear spoiler. While the base Stealth wears 15in tires, the midline models use 16s and the R/T 17s. Anti-lock brakes are standard on R/Ts and optional on the others. Topping off this success, Dodge was chosen to pace the 1991 Indianapolis 500 with a Stealth, and marked the occasion with a special Indy trim package. Stealths have been selling at a good clip, about 15,000 per year.

The 1993 Stealth R/T Turbo. This is usually a bad angle for a car, but this Dodge looks just as great from here as anywhere else.

Identification

Inimitable styling reminiscent of the best Italian supercars, especially those rear brake cooling ducts on lower body sides.

Appraisal

A Stealth feels more substantial than its Laser/Eclipse/Talon relatives and provides V-6 power, which elevates it from the Toyota Celica class into the Toyota Supra and Nissan 300ZX league. The R/T Turbo is terrifically fast, but not as fast as Dodge's claim of 0–60 in under five seconds; it's more like 6.5. Recent models still suffer from lack of refinement: turbo lag and a brutal ride. Stealths are geared way too high in order to avoid a federal gas guzzler tax—another proof positive that the feds screw up more than they fix. As a result, you're constantly forced to downshift, and the notchy, imprecise shifter is not entertaining.

Stealth interior folowed the best European sports car practice, with deep leather buckets and all controls within easy reach.

Summary and Prospects

Were it a convertible the Stealth would be instantly collectible; as it is, there are some questions to be resolved. One is, will all models be desirable or just the exotic, $30,000 R/T Turbo? My guess is the latter. The trouble is, the down-home models (which sold for as little as $16,000) look just like it, which won't help. But there's so much techno-think equipment packed in—electronic suspension, four-wheel drive, twin turbos, four-wheel steer—that the R/T Turbo just *has* to be collectible. Imagine having one twenty-five years from now, with all that stuff work-

On into the mid-1990s: the 1995 version. Stealth R/Ts are almost certainly assured of collector status in the 21st century. Remember, however, that R/T Turbos are low-volume items, under 2,000 a year.

ing (not to mention the adjustable exhaust, note: glass packs with an on-off switch). How to get one? There are two ways: visit your local Dodge dealer, bringing plenty wampum; or wait a few years and buy a good one off a used car lot. The problem with the former approach is that it takes a bundle; the problem with the latter is that R/T Turbos will have generally led hard lives.

Price History

Original list prices range between $15,000 and $35,000; all models are depreciating at present, though not at the rate of conventional cars.

Production

	1992	1993
(Sales)	16,926	14,556

Turbo R/T production was under 2,000 in both model years

Specifications

Engines

(net hp)

Type: overhead cam and double overhead cam V-6
3.0 liters, 181ci (3.59 x 2.99), 164hp (ohc, base model); 222hp (dohc, on ES and R/T); 300–320hp (R/T Turbo)

Chassis and Drivetrain

Unibody construction
Transmission: five-speed manual standard; automatic optional (not available on R/T Turbo)
Suspension: MacPherson struts with coil springs and anti-sway bar front; trailing-arm independent suspension with coil springs and sway bars rear

Measurements

Wheelbase (in): 97.2
Curb Weight (lbs): 3,076 (base); 3,175 (ES); 3,352 (R/T); 3,793 (R/T Turbo)
Length (in): 179.1

Performance

(R/T Turbo)
0–60mph (secs.): 6 in most tests.
Top speed (mph): 150
Mileage (mpg): 18–24

Chapter 32

Fun	Investment	Anguish
10	8	7

Viper
1992–

History

After a gestation period that would rival that of a brontosaurus, Dodge finally introduced its production supercar in early 1992, built by a tiny team of 120 to 160 "craftspersons" at the New Mack Avenue factory in east Detroit, which formerly built prototypes. A pure statement of classic sports car parameters in modern context, it was the work of a special team of eighty-five carefully screened Chrysler car nuts headed by Roy H. Sjoberg, who volunteered to develop it in an old AMC building that once housed Jeep engineering. Despite lengthy advance promotion, the Viper really didn't take long to create: it went from concept to production in thirty-six months, something of a record by Detroit standards. Of course it was a ground-up, name-your-price project, the kind that autoholics love. This shows in what they gave us.

The ultimate Dodge, an instantaneous

If Stealth knocked 'em flat, nobody believed the Viper was coming, except the car-of-the-month magazine crowd who warned for years that it was. Packing 400 net (yes) horsepower in its voluptuous resin transfer-molded plastic body, it takes the title as Cobra of the '90s. This is a factory photo of the 1995 model.

Viper for 1994: another angle on a beautiful brute. *Road & Track* appropriately summarizes the latest and greatest Dodge sporting car: "It has rekindled passions for an all-conquering, brawny-engine, front-midships roadster, passions that have been smoldering since the last 427 Shelby Cobras."

collectible, Viper is the best proof anyone could ask that Chrysler Corporation is back in business as the serious producer of serious motorcars. Under its resin-transfer-molded plastic panels is a thoroughbred mechanism spawned of competition and the hot dreams of carfolk: a welded tubular steel space frame, unequal-length A-arm suspension all round, 13in vented disc brakes, six speeds, and a pushrod V-10 displacing a full liter *more* than the vaunted Cobra 427. The Viper puts out 400 net horsepower; in the terms by which hp was measured in the Cobra's day, that would be close to 600 gross.

The engine is not only a first on a modern production car but a tremendously impressive powerplant compared to anything that has gone before. An all-aluminum V-10, its design is essentially the same as Chrysler's 360 V-8 with two extra cylinders, and robust construction, typified by its forged steel connecting rods and crankshaft. It develops prodigious torque, 450lb-ft at 3600rpm, and its torque curve is almost flat between 1500 and 5500rpm. Thanks to clever manifolding by consultant Lamborghini, it has the lowest coolant-temperature rise of any Chrysler engine ever built; being of all-aluminum construction, it is 100lb lighter than it would be with cast-iron heads and block.

Despite its heroic specifications, the Viper is a tractable car around town, requiring light clutch effort, short and easy throws of its Borg-Warner six-speed gearbox, and wrapping the driver in a simple, purposeful cockpit devoid of gimmicks. Unleashed, it goes like howling hell, corners on a dime. It's impractical as all get-out, politically incorrect, and socially unacceptable: in short, exactly the thing a real performance Dodge ought to be. Vipers are designed to be driven in the open air. There's a top and side curtains, but they look terrible and won't be seen often. Air conditioning, dealer-installed, is an option.

Identification

A two-seat roadster with targa top, the Viper is inimitable and unmistakable, bearing its own logos and fender identification. Resin transfer molding (RTM) body panels; only the lower front body enclosure is formed of sheet-molded compound, which is also found on the Corvette.

Appraisal

The best appraisal I've read from a collector's standpoint was by Ron Sessions in *Road & Track*: "The Viper isn't about numbers. It's about unbridled emotion on wheels. It's about explosive locomotion and the power to blast to 100 or 150mph at will without working up a sweat. It's about balance and a 50/50 weight distribution that lets a skilled driver arm-wrestle difficult corners, approach and dance on the edge of the laws of physics. . . . It is sending little ripples of excitement through the ranks of MoPar fans. The sort of excitement not experienced since the days of Hemi 'Cudas, 440 Six Pack Dodge Challengers and winged Charger Daytonas. It has rekindled passions for an all-conquering, brawny-engine, front-midships roadster, passions that have been smoldering since the last 427 Shelby Cobras."

Summary and Prospects

There's no need for conjecture about the Viper's collectibility. That's already assured. It costs $50,000 and though Dodge says production could reach 3,000 in 1994, it's doubtful that there will ever be much more volume than that. The nearest comparison to it as a product is the 1990 Corvette ZR-1, but there are differences in the Viper's favor: the ZR-1 was an option package whose main distinction was its exotic 32-valve V-8; the Viper is a ground-up creation that won't be duplicated or "prodified" in lower priced Dodges. Although speculators initially bid the ZR-1 ($70,000 sticker) to six figures, the bubble burst within the year, and it's now selling for well *below* sticker. But no discounts have been asked for the Viper, and it's into its third year now. On the other hand, do not pay more than sticker price for one, and expect to keep it a long time before prices start to turn upward. Also, the same money will probably buy you a good Charger R/T *and* a Challenger Six Pack. You pays your money and takes your choice.

Price History

Viper sticker prices are around $60,000. According to the CPI price guide, which only began listing used Vipers recently, the '92s are selling for $65,000–80,000 (which seems high) while the '93s (which saw much higher production) are at $53,500–66,000. That's virtually no depreciation, so far. If accurate, the reasons involve the rarity of the '92 and the very limited production of the '93. Exactly where the Viper will bottom out depends on how many they build and for how long. If they get down around $25,000 in three or four years, that might be the time to make your purchase decision.

Production

1992	1993
162	1,043

Specifications

Engines
(net hp)
Type: overhead valve pushrod aluminum V-10
8.0 liters, 488ci (4.00 x 3.88), 400hp

Chassis and Drivetrain
Fiberglass and aluminum body on tubular steel frame
Transmission: six-speed manual
Suspension: Upper and lower A-arms, coil springs, tube shocks and anti-sway bar front; upper and lower A-arms, toe links, coil springs, tube shocks and anti-sway bar rear

Measurements
Wheelbase (in): 96.2
Curb Weight (lbs): 3,500
Length (in): 175.1

Performance
0–60mph (secs.): 4.8
Standing 1/4 mile (secs. @ mph): 13.1 @ 109
Top speed (mph): 160
Mileage (mpg): 13/22, average 15

Index